Michelson-Morley Experiments

An Enigma for Physics and the History of Science

Michelson-Morley Experiments

Experiments

An Enigma for Physics and the History of Science

Maurizio Consoli

National Institute for Nuclear Physics, Italy

Alessandro Pluchino

University of Catania, Italy

World Scientific

NEW JERSEY · LONDON · SINGAPORE · BEIJING · SHANGHAI · HONG KONG · TAIPEI · CHENNAI · TOKYO

Published by

World Scientific Publishing Co. Pte. Ltd.

5 Toh Tuck Link, Singapore 596224

USA office: 27 Warren Street, Suite 401-402, Hackensack, NJ 07601

UK office: 57 Shelton Street, Covent Garden, London WC2H 9HE

British Library Cataloguing-in-Publication Data
A catalogue record for this book is available from the British Library.

The cover shows the scheme of the modern Michelson–Morley experiment by H. Müller *et al.* in *Phys. Rev. Lett.* 91 (2003) 020401 with the portraits of Albert Einstein and Hendrik Anton Lorentz.

MICHELSON–MORLEY EXPERIMENTS
An Enigma for Physics and the History of Science

Copyright © 2019 by World Scientific Publishing Co. Pte. Ltd.

ISBN 978-981-3278-18-9

For any available supplementary material, please visit
https://www.worldscientific.com/worldscibooks/10.1142/11209#t=suppl

Desk Editor: Christopher Teo

Typeset by Stallion Press
Email: enquiries@stallionpress.com

Preface

Subtle is the Lord but malicious He is not.

A. EINSTEIN, 1921

The Lord whose oracle is at Delphi
neither reveals nor conceals but gives a sign.

HERACLITUS, V Century B. C.

In 1887 Michelson and Morley tried to detect in laboratory a small difference of the velocity of light propagating in different directions that, according to classical physics, should have revealed the motion of the earth in the ether ("ether drift"). The result of their measurements, however, was much smaller than the classical prediction and considered as a typical instrumental artifact: a "null result". This was crucial to stimulate the first, pioneering formulations of the relativistic effects and, as such, represents a fundamental step in the history of science.

Nowadays, this original experiment and its early repetitions performed at the turn of 19th and 20th centuries (by Miller, Kennedy, Illingworth, Joos...) are considered as a venerable, well understood historical chapter for which, at least from a physical point of view, there is nothing more to refine or clarify. All emphasis is now on the modern versions of these experiments, with lasers stabilized by optical cavities that, apparently, have confirmed the null result by improving by many orders of magnitude on the limits placed by those original measurements.

Though, this is not necessarily true. In the original measurements, light was propagating in gaseous systems (air or helium at atmospheric pressure) while now, in modern experiments, light propagates in a high vacuum or inside solid dielectrics. Therefore, in principle, the difference with the modern experiments might not depend on the technological progress only but also on the different media that are tested thus preventing a straightforward comparison.

This is even more true if one takes into account that, in the past, greatest experts (as Hicks and Miller) have seriously questioned the traditional null interpretation of the very early measurements. The observed "fringe shifts", although much smaller than the predictions of classical physics, were often non negligible as compared to the extraordinary sensitivity of the interferometers. Therefore, in some alternative scheme, the small residuals could acquire a definite physical meaning.

By starting from this observation, in the last few years we have formulated a new theoretical framework where these residual effects could represent the first experimental indication for the earth motion within the Cosmic Microwave Background (CMB). In fact, in this alternative scheme, the small observed residuals show surprising correlations with the direct observations of the CMB dipole anisotropy with satellites in space.

The possibility of finally linking the CMB with the existence of a fundamental reference frame for relativity, and the substantial implications for the interpretation of non-locality in the quantum theory, would be of paramount importance. Therefore, we should preliminarily explain at least the key ingredients of our alternative scheme.

First of all, one should not compare the data with the classical predictions but impose that all measurable effects vanish exactly if the velocity of light c_γ propagating in the various interferometers, or more precisely its two-way combination $\bar{c}_\gamma$, coincides with the basic parameter c entering Lorentz transformations. This is the *ideal* vacuum limit of a refractive index $\mathcal{N} = 1$ where no ether drift should be observed. Instead if $\bar{c}_\gamma \neq c$, as for instance in the presence of matter, where light gets absorbed and then re-emitted, nothing would really prevent a non-zero light anisotropy $\Delta\bar{c}_\theta = \bar{c}_\gamma(\pi/2 + \theta) - \bar{c}_\gamma(\theta) \neq 0$.

Then, in the infinitesimal region $\mathcal{N} = 1 + \epsilon$, which corresponds for instance to gaseous systems, one can expand $\Delta\bar{c}_\theta$ in powers of the two small parameters ϵ and $\beta = v/c$, v being the velocity of the laboratory system with respect to the hypothetical preferred frame. By simple symmetry arguments, this expansion leads to the relation $\frac{|\Delta\bar{c}_\theta|}{c} \sim \epsilon\beta^2$ which is much smaller than the estimate $\frac{|\Delta \bar{c}_\theta|_{\text{class}}}{c} \sim \beta^2/2$ of the classical calculation. To have an idea, for experiments in air at room temperature and atmospheric pressure, where $\epsilon \sim 2.8 \cdot 10^{-4}$, and for the typical projection $v \sim 300$ km/s of the earth cosmic motion where $\beta^2 = 10^{-6}$, our estimate would still be about 17 times smaller than the classical prediction for the much smaller traditional orbital value $v = 30$ km/s where $\beta^2 = 10^{-8}$. For helium at room temperature and atmospheric pressure, where $\epsilon \sim 3.3 \cdot 10^{-5}$, our expectation would even be 150 times smaller. This could now explain the order of magnitude of the observed effects.

The other peculiar aspect of our analysis concerns the time dependence of the data. Here, the traditional view is that, for short-time observations of a few days, where there are no sizeable changes in the orbital motion, a genuine physical signal should precisely follow the slow and regular modulations induced by the earth rotation. The fringe shifts instead were showing an irregular behavior indicating sizeably different directions of the drift at the same hour on consecutive days so that statistical averages were much smaller than all individual values. Within the traditional view, this has always represented a strong argument to interpret the measurements as mere instrumental artifacts.

Again, however, there might be a logical gap. The relation between the macroscopic earth motion and the microscopic propagation of light in a laboratory depends on a complicated chain of effects and, ultimately, on the physical nature of the vacuum. By comparing with the motion of a body in a fluid, the standard view corresponds to a form of regular, laminar flow where global and local velocity fields coincide. Instead, some arguments suggest that the *physical vacuum* might rather behave as a stochastic medium which resembles a highly turbulent fluid where large-scale and small-scale flows are only *indirectly* related.

In this different perspective, with forms of turbulence which, as in most models, become statistically isotropic at small scales, the direction of the local drift is a completely random quantity that has no definite limit by combining a large number of observations. Thus, one should first analyze the data in phase and amplitude (which give respectively the instantaneous direction and magnitude of the drift) and then concentrate on the latter which is a positive-definite quantity and remains non-zero under any averaging procedure. In this alternative picture, a non-vanishing amplitude (i.e. definitely larger than the experimental resolution) is the signature to separate an irregular, but genuine, signal from instrumental noise.

By implementing these two ingredients, the classical experiments in gaseous systems can now become consistent with the earth velocity of 370 km/s deduced from the direct CMB observations. In particular, from a fit to Joos's 1930 very precise observations (data collected during all 24 hours to cover the full sidereal day and recorded automatically by photocamera), we have also obtained some information on the angular parameters of the earth motion, namely right ascension $\alpha(\text{fit} - \text{Joos}) = (168\pm30)$ degrees and angular declination $\gamma(\text{fit} - \text{Joos}) = (-13\pm14)$ degrees, to compare with the present values $\alpha(\text{CMB}) \sim 168$ degrees and $\gamma(\text{CMB}) \sim -7$ degrees. This consistency gives good motivations for a new generation of dedicated experiments to reproduce the experimental conditions of those old measurements with today's much greater accuracy.

Meanwhile, waiting for this definitive test, we have tried to obtain a different check with modern experiments *in vacuum*. The point is that in the *physical vacuum* the velocity of light may still differ from the parameter c of Lorentz transformations. This might be due to several reasons. For instance, some authors have suggested that the curvature observed in a gravitational field might represent a phenomenon which emerges from a fundamentally flat space-time. This would be in analogy with some condensed-matter systems (such as moving fluids, Bose-Einstein condensates...) at length scales much larger than the size of their elementary constituents. In this picture, one expects a tiny vacuum refractivity $\epsilon_v \sim 10^{-9}$ which accounts for the difference between an apparatus in an ideal freely-falling frame and an apparatus on the earth surface.

Then, if our interpretation of the classical experiments is correct, we would also expect a very small anisotropy $\frac{|\Delta \bar{c}_\theta|_v}{c} \sim \epsilon_v \beta^2 \sim 10^{-15}$ which could be detected by measuring the frequency shift of two vacuum optical resonators. More precisely, in our picture, this is the expected magnitude of the *instantaneous*, irregular signal. Its statistical average $\frac{|\langle \Delta \bar{c}_\theta \rangle_v|}{c}$ after many observations should instead be much smaller, say 10^{-18}, 10^{-19}..., and vanish in the limit of an infinite statistics. As we will illustrate, this expectation is consistent with the most recent room temperature and cryogenic vacuum experiments thus providing further support for our alternative interpretation.

Now, as it is well known, symmetry arguments give often a good description of phenomena independently of the underlying physical mechanisms. As such, our view of the classical experiments in gaseous systems, in terms of a light anisotropy $\frac{|\Delta \bar{c}_\theta|}{c} \sim \epsilon \beta^2$, does not necessarily contradict the standard interpretation of those old measurements as due to thermal disturbances. Indeed, these disturbances are also known to become smaller and smaller when $\epsilon \to 0$.

For this reason, and for the overall consistency of the data, the small temperature variations of a millikelvin in the air of the optical arms assumed by various authors (and never fully understood) to explain Miller's Mt. Wilson observations might have a *non-local* origin somehow associated with an absolute earth velocity v. After all, our motion within the CMB gives the same order of magnitude $[\Delta T(\theta)]_{\mathrm{CMB}} \sim \pm 3$ mK. As we will show, this thermal interpretation could provide a dynamical basis for the *enhancement* found in the gas case (i.e. the observed magnitudes $\frac{|\Delta \bar{c}_\theta|_{\mathrm{air}}}{c} = \mathcal{O}(10^{-10})$ for air and $\frac{|\Delta \bar{c}_\theta|_{\mathrm{helium}}}{c} = \mathcal{O}(10^{-11})$ for gaseous helium vs. the much smaller vacuum value $\frac{|\Delta \bar{c}_\theta|_v}{c} \lesssim 10^{-15}$) and, at the same time, could also help to understand the differences and the analogies with the most precise experiment in solid dielectrics where again an *instantaneous* value $\frac{|\Delta \bar{c}_\theta|_{\mathrm{solid}}}{c} \lesssim 10^{-15}$ (as in the vacuum case) is presently observed. In this way, symmetry arguments, on the one hand, would motivate and, on the other hand, would find justification in underlying physical mechanisms, with an overall increase of our understanding.

We emphasize that this book is primarily a monograph about the *physics* of these experiments. However, the *history* of this research is also interesting and sometimes even dramatic for the strong personal commitment of some scientist. For this reason, several historical accounts have been included as a useful supplementary material.

In conclusion, our work should motivate the reader to sharpen his own understanding of both classical and modern Michelson-Morley experiments. Then, it will become evident that their standard null interpretation, presented in all textbooks and specialized reviews as the most evident scientific truth, is very far from obvious and most probably wrong. This is why these experiments represent an enigma for physics and the history of science. In view of their fundamental importance, we hope that our book will induce to refine substantially the experimental tests and the analysis of the data thus contributing to reach a higher level of collective awareness.

Acknowledgments

We thank our friends and colleagues Evelina Costanzo, Caroline Matheson, Angelo Pagano, Lorenzo Pappalardo and Andrea Rapisarda for many useful discussions and their precious collaboration.

Contents

Chapter 1

You, my honored Herr Michelson began this work when I
was only a small boy, not even a meter high. It was you
who led the physicists into new paths, and through your
marvelous experimental labors prepared for the development
of the relativity theory. You uncovered a dangerous
weakness in the ether theory of light as it then existed,
and stimulated the thoughts of H. A. Lorentz and Fitzgerald
from which the special theory of relativity emerged.

A. EINSTEIN, Speech in honor of Michelson, Pasadena,15 January 1931.

1.1 Premise

The Michelson-Morley experiment [1] was designed to check Maxwell's classical prediction [2] that if the earth drifts in the ether with a velocity v there should be an anisotropy $\frac{|\Delta \bar{c}_\theta|}{c} \sim \frac{v^2}{c^2}$ of the two-way velocity of light in the earth frame[1]. Michelson's idea was to detect this tiny effect by observing the interference fringes of two light rays propagating back and forth along perpendicular directions.

To introduce the argument, let us consider the two-way velocity of light $\bar{c}_\gamma(\theta)$. This is the only one that can be measured unambiguously and is

---

[1] Actually, Maxwell's estimate is larger by a factor of two than the standard classical prediction $\frac{|\Delta \bar{c}_\theta|}{c} \sim \frac{v^2}{2c^2}$, see Chapt.3.

1

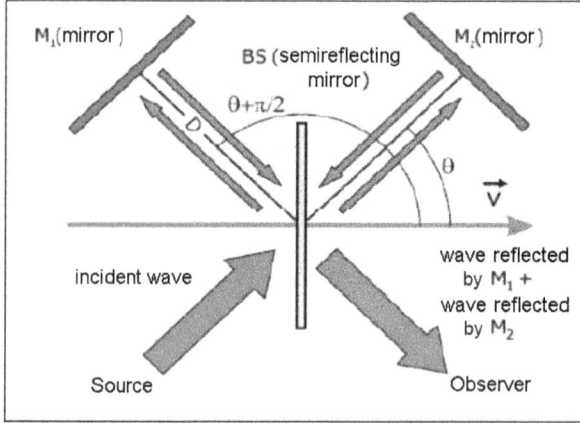

Fig. 1.1   *The typical scheme of Michelson's interferometer.*

defined in terms of the one-way velocity $c_\gamma(\theta)$ as

$$\bar{c}_\gamma(\theta) = \frac{2\, c_\gamma(\theta) c_\gamma(\pi + \theta)}{c_\gamma(\theta) + c_\gamma(\pi + \theta)} \tag{1.1}$$

where θ represents the angle between the direction of light propagation and the earth velocity with respect to the hypothetical preferred frame Σ.

By introducing the anisotropy

$$\Delta\bar{c}_\theta = \bar{c}_\gamma(\pi/2 + \theta) - \bar{c}_\gamma(\theta) \tag{1.2}$$

there is a simple relation with the time difference $\Delta t(\theta)$ for light propagation back and forth along perpendicular rods of length D (see Fig.1.1)

$$\Delta t(\theta) = \frac{2D}{\bar{c}_\gamma(\theta)} - \frac{2D}{\bar{c}_\gamma(\pi/2 + \theta)} \sim \frac{2D}{c} \frac{\Delta\bar{c}_\theta}{c} \tag{1.3}$$

(where, in the last relation, we have assumed that light propagates in a medium of refractive index $\mathcal{N} = 1 + \epsilon$, with $\epsilon \ll 1$). This gives the fringe patterns (λ is the light wavelength)

$$\frac{\Delta\lambda(\theta)}{\lambda} \sim \frac{2D}{\lambda} \frac{\Delta\bar{c}_\theta}{c} \tag{1.4}$$

and could be measured, in principle, by rotating the apparatus.

The classical prediction (see e.g. [3] for a simple derivation) was

$$\left[\frac{\Delta\lambda(\theta)}{\lambda}\right]_{\text{class}} \sim \frac{D}{\lambda} \frac{v^2}{c^2} \cos 2\theta \tag{1.5}$$

and, for the Michelson-Morley apparatus, the relevant value was $(D/\lambda) \sim 2 \cdot 10^7$. Therefore, for $v = 30$ km/s (the earth orbital velocity about the sun, and consequently the minimum anticipated drift velocity) where $v^2/c^2 = 10^{-8}$, under a 90 degree rotation, one was expecting a shift

$$\left[\frac{\Delta\lambda(0)}{\lambda} - \frac{\Delta\lambda(\pi/2)}{\lambda}\right]_{\text{class}} \sim \frac{2D}{\lambda}\frac{v^2}{c^2} \sim 0.4 \qquad (1.6)$$

that would have been about *hundred times larger* than the extraordinary sensitivity of the apparatus, about ± 0.004 [1, 4, 5].

Instead, in the various experimental sessions, the observed shifts were about $10 \div 20$ times smaller than expected [6, 7]. By using Eq.(1.5), these values were indicating earth velocities of about $6 \div 10$ km/s which have no obvious interpretation. In addition, the observed pattern was irregular because observations performed at the same hour on consecutive days were showing sizeable differences. The simultaneous presence of these two aspects gave a strong argument to consider the data as typical instrumental effects, i.e. a "null" result.

The acceptance of this view, indicating a failure of the classical ideas and/or the non-existence of the ether, had a strong impact on the scientific ambiance and was crucial to stimulate the first, pioneering formulations of the relativistic length contraction and time dilation effects, by Fitzgerald in 1889 [8], Lorentz in 1895 [9] and 1899 [10], Larmor in 1897 [11] and 1900 [12]. These original developments of the theory of the electromagnetic ether induced Lorentz in 1904 [13] and Poincaré in 1905 [14] to derive a particular set of transformations of the space-time coordinates (Lorentz Transformations) : "Applying one of such transformations amounts to an overall translation to the whole system. Then two frames, one at rest in the ether and one in uniform translation, become the perfect images of each other". This statement of Poincaré in 1905 was the precise formalization of the Principle of Relativity, already proposed by him in La Science et l'Hypothese (Flammarion, Paris 1902) and at the 1904 St. Louis Conference [15][2].

---

[2]Poincaré's precise words to formulate this principle in his 1904 address are: "The principle of relativity, according to which the laws of physical phenomena should be the same, whether for an observer fixed, or for an observer carried along in a uniform movement of translation; so that we have not and could not have any means of discerning whether or not we are carried along in such a motion" [16]. In the same address, Poincaré was also concluding that "From all these results, if they are confirmed, would arise an entirely new mechanics, which would be, above all, characterized by this fact, that no velocity could surpass that of light, any more than any temperature could fall below the zero absolute, because bodies would oppose an increasing inertia to the causes, which would tend to accelerate their motion; and this inertia would become infinite when one approached the velocity of light" [16].

This first historical phase and its relation with special relativity [17] can be well described by quoting, twice, Einstein himself. The first quotation is from his address to Michelson during a social gathering of scientists at the California Institute of Technology in mid-January 1931: "You, my honored Herr Michelson began this work when I was only a small boy, not even a meter high. It was you who led the physicists into new paths, and through your marvelous experimental labors prepared for the development of the relativity theory. You uncovered a dangerous weakness in the ether theory of light as it then existed, and stimulated the thoughts of H. A. Lorentz and Fitzgerald from which the special theory of relativity emerged" [18].

The second quotation is from an Einstein's interview delivered in 1955, a few months before his death. When asked, once more, about his original view and the relation with previous work he said: " There is no doubt that the theory of relativity, if we regard its development in retrospect, was ripe for discovery in 1905. Lorentz had already observed that the transformations which later were known by his name were essential for the analysis of Maxwell equations and Poincaré had even penetrated deeper into these connections. Concerning myself, I knew only Lorentz' important work of 1895 but not his later work nor the consecutive investigations by Poincaré. In this sense my work of 1905 was independent. The new feature of it was the realization that the bearing of Lorentz transformations transcended its connection with Maxwell equations and was concerned with the nature of space and time in general. The new result was that Lorentz invariance was the general condition for any physical theory" [19].

Thus, one could summarize as follows: (i) the Michelson-Morley experiment was crucial for the first formulation of the relativistic effects within the theory of the electromagnetic ether (ii) later on, by Einstein, relativity was recognized as a doctrine of nature and formulated in an axiomatic form, free of any association with ether and electromagnetism[3].

---

[3]Over the years, Einstein made different statements about the inception of relativity and the possible influence that the Michelson-Morley experiment had on his views. For instance, Holton [5] reports the following sentence: "In my own development Michelson's result has not had a considerable influence. I even do not remember if I knew of it at all when I wrote my first paper on the subject (1905). The explanation is that I was, for general reasons, firmly convinced that there does not exist absolute motion and my problem was only how this could be reconciled with our knowledge of electrodynamics. One can therefore understand why in my personal struggle Michelson's experiment played no role or at least no decisive role". At the same time, this other statement is reported by van Dongen [20]: "As a young man I was interested, as a physicist, in the question what is the nature of light, and, in particular, what is the nature of light with respect to bodies. That is, as a child I was already taught that light is subordinate to the oscillations of the

Such premise is essential to properly frame the Michelson-Morley experiment in the history of physics. At the same time, nowadays, there is the tendency to consider this fundamental experiment, and its classical repetitions at the beginning of 20th century by Miller [7], Illingworth [21], Joos [22] ... as an old, well understood historical chapter for which there is nothing more to refine or clarify. All emphasis is now on the modern versions of these experiments, with lasers stabilized by optical cavities (see e.g. [23] for a review), which apparently have improved by orders of magnitude on those original measurements [24].

However, a basic aspect has been overlooked by most authors. The various measurements were performed in different conditions, i.e. with light propagating in gaseous media (as in [1, 7, 21, 22]) or in a high vacuum (as in [25–27]) or inside dielectrics with a large refractive index (as in [24, 30]) and there could be physical reasons which prevent a straightforward comparison. In this case, the difference between old experiments (in gases) and modern experiments (in vacuum or solid dielectrics) might not depend on the technological progress only but also on the different media that were tested. Then, if the small residuals of those original experiments were not mere instrumental artifacts, there would be substantial implications for both physics and history of science.

1.2   Lorentz vs. Einstein

Before going deeper into the analysis of the experiments, we want to add some general comment about Lorentz' and Einstein's views of relativity. Apart from all historical aspects, the basic difference could simply be phrased as follows. In a "Lorentzian" approach, the relativistic effects originate from the *individual* motion of each observer S', S"...with respect to some preferred reference frame Σ, a convenient redefinition of Lorentz' ether. Instead, according to Einstein, eliminating the concept of the ether as a preferred frame leads to interpret the same effects as consequences of the *relative* motion of each pair of observers S' and S".

---

light ether. If that is the case, then one should be able to detect it, and thus I thought about whether it would be possible to perceive through some experiment that the earth moves in the ether. But when I was a student, I saw that experiments of this kind had already been made, in particular by your compatriot, Michelson. He proved that one does not notice anything on earth that it moves, but that everything takes place on earth as if the earth is in a state of rest". In spite of these contradictions, we trust in our synthesis (points i) and ii) above) which derives from two consistent Einstein's citations and fits well with the historical evolution of the scientific ideas.

In spite of this difference, it is generally assumed that there is a substantial phenomenological equivalence between the two formulations. This point of view was, for instance, already clearly expressed by Ehrenfest in his lecture 'On the crisis of the light ether hypothesis' (Leyden, December 1912) as follows: "So, we see that the ether-less theory of Einstein demands exactly the same here as the ether theory of Lorentz. It is, in fact, because of this circumstance, that according to Einstein's theory an observer must observe exactly the same contractions, changes of rate, etc. in the measuring rods, clocks, etc. moving with respect to him as in the Lorentzian theory. And let it be said here right away and in all generality. As a matter of principle, there is no experimentum crucis between the two theories". In fact, independently of all interpretative aspects, the basic quantitative ingredients, namely Lorentz transformations, are the same in both formulations.

Then, one may get the impression that the present supremacy of Einstein's interpretation depends on the null interpretation of the ether-drift experiments where one attempts to measure the "absolute" earth velocity with respect to the hypothetical Σ. Though, this is not true. In a Lorentzian perspective, if the velocity of light c_γ propagating in the various interferometers coincides with the basic parameter c entering Lorentz transformations, relativistic effects conspire to make undetectable the individual velocity parameter of each observer. For this reason, a null result of the ether-drift experiments should *not* automatically be interpreted as a confirmation of special relativity. The motion with respect to Σ might well remain unobservable, yet one could interpret relativity 'á la Lorentz'. As emphasized by Bell [31], see also [32, 33], this change of perspective, which may be useful for pedagogical reasons, could also be crucial to reconcile faster-than-light signals with causality [34, 35] and thus provide a very different view of the apparent non-local aspects of the quantum theory [36].

In our context of the ether-drift experiments, we want to add one more remark about Einstein's and Lorentz' points of views. According to Einstein, the original hypothesis of a real, physical length contraction along the direction of motion, to compensate the difference of the light velocity and thus explain the failure of the classical relation (1.5), was important from a historical point of view but highly unsatisfactory as ultimate explanation. For him, it was preferable to build the theory by postulating the impossibility-in-principle of discovering an absolute state of motion, or equivalently of assigning "a velocity vector to a point in empty space where electromagnetic processes take place" [17]. For Lorentz, on the other hand, only a conspiracy of effects, associated with the equality $c_\gamma = c$, was preventing to detect the motion with respect to the ether which, however different might be from ordinary matter, is nevertheless endowed with a certain degree of substantiality. For this reason, in his view, "it seems natural not to assume at starting that it can never make any difference whether a body moves through the ether or not" [37].

Adopting such a "Lorentzian" open mind was for us an important motivation to undertake a re-analysis [38] of the early ether-drift experiments in gaseous media and check the claims of those experts [6,7] that over the years have seriously questioned the standard null interpretation. In their opinion, in fact, the fringe shifts were much smaller than the classical predictions but not always negligible as compared to the extraordinary sensitivity of the interferometers. This means that, in some alternative model, the small residuals can acquire a definite physical meaning.

1.3 Classical ether-drift experiments: Just null results?

For our analysis of the ether-drift experiments, we shall rely on two basic assumptions, namely (i) the existence of a preferred frame Σ where light propagation is seen isotropic and (ii) the validity of Lorentz transformations. This means that any anisotropy in the earth frame S' should vanish identically either when the earth velocity $v = 0$ (i.e. $S' \equiv \Sigma$) or when $\mathcal{N} = 1$, i.e. when $c_\gamma \equiv c^4$. The reason for the substantial suppression of the fringe shifts can then be understood by exploring the possible functional forms for the two-way velocity of light in the limit of refractive index $\mathcal{N} = 1 + \epsilon$. With our premise, in fact, for $\epsilon \ll 1$, one can expand in powers of ϵ and $\beta = v/c$ and it is elementary to show, see [38,40] and the following Chap.6, that the leading term for a possible anisotropy of the two-way velocity of light is

$$\frac{\Delta \bar{c}_\theta}{c} \sim \epsilon \beta^2 \cos 2\theta. \tag{1.7}$$

Therefore, the fringe shifts are now predicted

$$\frac{\Delta \lambda(\theta)}{\lambda} \sim \frac{D}{\lambda} \frac{2\epsilon v^2}{c^2} \cos 2\theta \tag{1.8}$$

and are suppressed by the tiny factor 2ϵ with respect to Eq.(1.5). As such, this basic difference can be reabsorbed into an *observable* velocity

$$v_{\text{obs}}^2 \sim 2\epsilon v^2 \tag{1.9}$$

---

[4]Actually, De Abreu and Guerra have shown [39] that the null result of a Michelson-Morley experiment in an ideal vacuum can be deduced without using Lorentz transformations, but only from general assumptions on the choice of the admissible clocks.

which depends on the refractive index and is the one traditionally reported in the classical analysis of the data.

More precisely, one can *define* the observable velocity v_{obs} through the relation

$$\frac{\Delta\lambda(\theta)}{\lambda} \sim \frac{D}{\lambda} \frac{v_{obs}^2}{c^2} \cos 2\theta \qquad (1.10)$$

to make clear that this is the velocity extracted from the fringe shifts. In classical physics one has $v_{obs} = v$, where v is the *kinematical* velocity. However, in other contexts the two quantities are different.

In any case, due to the formal properties of the two way velocity of light, the fringe shifts should represent a 2nd-harmonic effect, i.e. periodic in θ in the range $[0, \pi]$, with amplitude

$$A_2 \sim \frac{D}{\lambda} \frac{v_{obs}^2}{c^2}. \qquad (1.11)$$

Notice also that in the classical experiments, where one was always assuming the identity $v_{obs} = v$, it was customary to formulate predictions for the orbital earth velocity $v = 30$ km/s. For this reason, we will often refer to the classical expected amplitude

$$A_2^{class} \sim \frac{D}{\lambda} \frac{(30 \text{ km/s})^2}{c^2} \qquad (1.12)$$

as a convenient reference value to compute the observable velocity from the experimental amplitude A_2^{EXP} through the relation

$$v_{obs} \sim 30 \text{ km/s} \sqrt{\frac{A_2^{EXP}}{A_2^{class}}}. \qquad (1.13)$$

The main conclusion of this discussion is that those small *observable* velocities obtained from the measured fringe shifts, typically $v_{obs} = 6 \div 10$ km/s for experiments in air (where ϵ is about $2.8 \cdot 10^{-4}$) and $v_{obs} = 2 \div 3$ km/s for experiments in helium (where ϵ is about $3.3 \cdot 10^{-5}$), can now become consistent with the average *kinematical* earth velocity $v \sim 370$ km/s obtained by studying, with aircraft and satellites, the Cosmic Microwave Background (CMB).

Apart from the order of magnitude of the fringe shifts, another important aspect concerns the time dependence of the data. Traditionally, it has been always assumed that, for short-time observations of a few days, where there are no sizeable changes in the earth orbital velocity, the time dependence of a genuine physical signal should reproduce the slow and regular modulations induced by the earth rotation. The data instead, for both classical and modern experiments, have always shown a very irregular behavior. As a consequence, all statistical averages are much smaller than the instantaneous values. This difference, between individual measurements and statistical averages, has always represented a strong argument to interpret the data as mere instrumental artifacts.

However, again, could there be an alternative interpretative scheme? A possibility, would be to characterize the signal as in the standard simulations adopted to model turbulent flows. This idea derives from realizing that, at the most fundamental level, light propagation (e.g. inside an optical cavity) takes place in that substratum which we could call *physical vacuum*[5]. This is dragged along the earth motion but, so to speak, is not rigidly connected with the solid parts of the apparatus as fixed in the laboratory. Therefore, if one would try to characterize its local state of motion, say $v_\mu(t)$, this does not necessarily coincide with the projection of the global earth motion, say $\tilde{v}_\mu(t)$, at the observation site. The latter is a smooth function while the former, $v_\mu(t)$, in principle is unknown. By comparing with the motion of a body in a fluid, the equality $v_\mu(t) = \tilde{v}_\mu(t)$ amounts to assume a form of regular, laminar flow where global and local velocity fields coincide.

Though, this is not necessarily true. For instance, it would be natural to compare the physical vacuum to a fluid with vanishing viscosity (or infinite Reynolds number for any flow velocity). But, within the framework of the Navier-Stokes equation, the picture of a laminar flow is by no means obvious due to the subtlety of the zero-viscosity limit, see for instance the discussion given by Feynman in Sec. 41.5, Vol.II of his Lectures [41]. The reason is that the velocity field of such a hypothetical fluid cannot be a differentiable function [42]. Instead, one should think in terms of a continuous, nowhere differentiable function[6], similar to an ideal Brownian path [43]. This leads to the idea of the vacuum as a fundamental stochastic medium, somehow similar to a highly turbulent fluid, consistently with some basic foundational aspects of both quantum physics and relativity [45].

For these reasons, it becomes conceivable that, as in turbulent flows, the local $v_\mu(t)$ exhibits random fluctuations while the global $\tilde{v}_\mu(t)$ just determines its typical limiting boundaries. Although the random $v_\mu(t)$ cannot

---

[5]Maxwell's original argument in favor of an ether was indeed considering this basic aspect of light propagation [2].

[6]Onsager's argument relies on the impossibility, in the zero-viscosity limit, to satisfy the inequality $|\mathbf{v}(\mathbf{x}+\mathbf{l}) - \mathbf{v}(\mathbf{x})| < (\text{const.})l^n$, with $n > 1/3$. Kolmogorov's theory [44] corresponds to $n = 1/3$.

be computed exactly, one could still estimate its statistical properties by numerical simulations [38, 45]. To this end, one could start by assuming forms of turbulence or intermittency which, as it is generally accepted in the limit of zero viscosity, become statistically isotropic at small scales. In this way, see the following Chap.6, one could easily explain the irregular character of the data because, whatever the macroscopic earth motion, the average of all vectorial quantities (such as the fringe shifts at the various angles) would tend to zero by increasing more and more the statistics. In this framework, is not surprising that from instantaneous measurements of given magnitude one ends up with smaller and smaller statistical averages. This trend, by itself, might not imply that there is no physical signal.

1.4   A universal thermal gradient, the CMB and the vacuum structure

Now, it is well known that symmetry arguments, of the type used to derive Eq.(1.7), can provide a successful description of phenomena *independently* of the particular dynamics. Still, one may wonder about the underlying physical mechanisms. Namely, when considering light propagation in a gas, why there should be an anisotropy in the earth laboratory where (the container of) the gas is at rest?

For instance, a possibility is that the electromagnetic field of the incoming radiation produces different polarizations in different directions depending on the state of motion of the medium. Such mechanism should act in both weakly bound gaseous matter and strongly bound solid dielectrics with the final conclusion that light anisotropy would increase proportionally to the refractivity of the medium. This is in contrast with the result of the Shamir-Fox [30] experiment in perspex where no particular enhancement was observed (with respect to the Michelson-Morley experiment in air).

As an alternative possibility, it was noted in [38, 40] (see also Chap.6) that the trend in Eq.(1.7) is just a special case of a more general structure where light anisotropy originates from convective currents of the gas molecules, along the optical paths, associated with an absolute velocity v. Therefore, on the basis of the traditional thermal interpretation[7] of the residuals, and of the consistency of the kinematical velocities obtained

---

[7]The idea that temperature differences along the optical paths could be crucial dates back to Helmholtz, already at the time of the first 1881 Michelson [46] experiment in Potsdam. The same emphasis on temperature differences is also found in the critical reanalysis of Miller's observations performed by Shankland et al. [47].

from measurements in different laboratories and different conditions, it becomes conceivable that such universal effect in gaseous systems reflects the existence of a *non-local* thermal gradient.

At the beginning, the ultimate explanation for the sought non-local thermal gradient was searched for [38, 40] in the fundamental energy flow which, on the basis of general arguments, is expected in a quantum vacuum which is not exactly Lorentz invariant and thus sets a fundamental preferred reference frame (see Chap.6).

Later on, however, it was argued [48] that the required physical mechanism could perhaps be related to the temperature variations associated with the CMB *kinematic dipole* [49–51] . This is interpreted as a Doppler effect due to a motion of the solar system with average velocity $v \sim 370$ km/s toward a point in the sky of right ascension $\alpha \sim 168^o$ and declination $\gamma \sim -7^o$ and produces angular variations of a few millikelvin which would fit well with the typical magnitude of the periodic temperature differences in the air of the optical arms, about $(1 \div 2)$ mK [47,52], which in principle could explain away the typical fringe shifts observed by Miller at Mount Wilson[8].

To check this interpretation, a new generation of dedicated experiments is needed to reproduce the experimental conditions of those early measurements with today's much greater accuracy. The essential ingredient is that the optical resonators which nowadays are coupled to the lasers should be filled by gaseous media. Such experiments would be along the lines of ref. [53] where just the use of optical cavities filled with different forms of matter was considered as a useful complementary tool to study deviations from exact Lorentz invariance.

In these modern ether-drift experiments one looks for a possible anisotropy of the two-way velocity of light through the relative frequency shift $\Delta\nu(\theta)$ of two orthogonal optical resonators (for a review see e.g. [23]). In units of their natural frequency ν_0, we thus predict a frequency shift

$$\left[\frac{\Delta\nu(\theta)}{\nu_0}\right]_{\text{gas}} \sim \left[\frac{\Delta\bar{c}_\theta}{c}\right]_{\text{gas}} \sim (\mathcal{N}_{\text{gas}} - 1)\,(v^2/c^2)\cos 2\theta. \qquad (1.14)$$

---

[8]We emphasize that our interpretation of light anisotropy $\frac{\Delta\bar{c}_\theta}{c} \sim \epsilon v^2/c^2$ in gaseous systems, as originating from a non-local thermal gradient, applies to *all* classical ether-drift experiments and not just to Miller's Mount Wilson measurements. This is an important difference with the standard point of view, deriving from the article of the Shankland team [47], which tends to distinguish sharply Miller's observations from all other experiments. These aspects will be discussed in great detail in the following Chaps.5, 6 and 7.

This type of thermal interpretation would also explain why the same trend $\frac{\Delta \bar{c}_\theta}{c} \sim (\mathcal{N} - 1)v^2/c^2$ does *not* extend to experiments in solid dielectrics, where the refractivity $\mathcal{N} - 1$ is of order unity, as with the mentioned Shamir-Fox [30] experiment in perspex ($\mathcal{N} = 1.5$). In this case, in fact, a small temperature gradient would mainly dissipate by heat conduction without generating any appreciable particle motion or light anisotropy in the rest frame of the apparatus. Hence, the non-trivial physical difference between classical experiments (in gaseous systems) and modern experiments (in a very high vacuum or in solid dielectrics).

Now, conceptually, explaining the residuals of the classical ether-drift experiments as non-local thermal effects, eventually related to the CMB temperature dipole, is different from introducing a preferred frame through a Lorentz-non-invariant vacuum state. To try to disentangle the two mechanisms, we have thus started to look at experiments where optical cavities are maintained in an extremely high vacuum, both at room temperature and in the cryogenic regime. The reason is that, in this limit, where any residual gaseous matter is totally negligible, a temperature gradient of a millikelvin cannot produce any observable light anisotropy. Therefore, if some infinitesimal effect still persists, the idea of a fundamental preferred frame would find additional support.

As a definite scenario to analyze these experiments, one can again consider the same scheme where $\bar{c}_\gamma \neq c$ so that the ideal equality $\mathcal{N} = 1$ does not hold exactly in the *physical* vacuum. As a possible motivation, it was proposed [57] that an effective vacuum value $\mathcal{N}_v = 1 + \epsilon_v$, with $\epsilon_v \sim 10^{-9}$, could reveal the different refractivity between an apparatus in an ideal freely-falling frame and an apparatus on the earth surface. This difference is expected if the curvature observed in a gravitational field is an emergent phenomenon from a fundamentally flat space-time. Then, the existence of a preferred frame would imply in our picture a definite, instantaneous $\frac{|\Delta \bar{c}_\theta|}{c} \sim \epsilon_v \beta^2 \sim 10^{-15}$ which coexists with a much smaller statistical average $|\langle \frac{\Delta \bar{c}_\theta}{c} \rangle| \ll 10^{-15}$. By assuming the same form of cosmic motion as for the classical experiments, in the following Chap.7, this expectation will be shown to be consistent with our numerical simulations of the most recent room temperature and cryogenic vacuum experiments.

Likewise, the existence of a fundamental 10^{-15} instantaneous signal, with very precise measurements, should also show up in solid dielectrics where, as anticipated, there should be no particular enhancement with respect to the vacuum case. This expectation is consistent with the cryogenic experiment of ref. [24] where most electromagnetic energy propagates in a solid with a refractive index $\mathcal{N} \sim 3$ (at microwave frequencies) but again a 10^{-15} instantaneous signal is observed.

To conclude this introductive chapter, we emphasize that, apart from its relevance for our view of relativity and for the history of science, a check of our predictions could have other non-trivial implications. In fact, suppose some future experiment would confirm the unambiguous detection of a universal signal in gaseous systems as in Eq.(1.14). In our interpretation, this would mean that light propagation in gaseous systems is modified due to a non-local temperature gradient, somehow associated with our motion within the CMB. But, of course, all physical systems on the moving earth would also be exposed to the same energy flow. This is very weak today but was substantially larger in the past when the temperature of the CMB was much higher. For this reason, one may speculate [58] on the role that this gradient might have played for the chemistry of liquid water. More in general, it is known [59, 60] that an external energy flow can induce forms of spontaneous self-organization in matter. In this sense, a universal thermal gradient could increase the efficiency of physical systems and provide a microscopic, dynamical mechanism to produce those macroscopic aspects (self-organized criticality, large-scale fluctuations, fat-tailed probability density functions...) which characterize the behavior of many complex systems, see e.g. [61–65] .

In the following, we will start in Chap.2 with some historical accounts on the ether conceptions that finally, at the end of XIX century, gave the motivation for the ether-drift experiments.

In Chap.3, we will concentrate on Michelson, on his early attempts to measure an ether-drift and on the original Michelson-Morley experiment.

In Chap.4, we will briefly comment on the inception of relativity and on its implications for the analysis of the experiments.

In Chap.5 we will report on the early repetitions of the Michelson-Morley experiment and on their traditional interpretation.

In Chap.6, we will re-consider the whole issue from scratch by introducing a modern formalism to re-interpret both the classical ether-drift experiments and the modern versions with optical resonators.

Finally, in Chap.7, we will extend our analysis to the present experiments in vacuum and solid dielectrics by discussing in more detail the various aspects briefly illustrated above.

Chapter 2

There exists a matter, distributed in the whole universe as a continuum, uniformly penetrating all bodies and filling all spaces, be it called ether, caloric or whatever, which is not a hypothetical material (for the purpose of explaining certain phenomena, and more or less conjuring up causes for given effects); rather it can be recognized and postulated "a priori" as an element necessarily belonging to the transition from the metaphysical foundation of natural science to physics.

I. KANT, Opus Postumum.

2.1 Some historical notes on the (a)ether

To better appreciate the 19th century ether conception, which was at the base of the first ether-drift experiments, we shall briefly sketch in this chapter the views of some great philosophers and scientists. For a more complete information, the reader may consult, for instance, Whittaker's book [66] and Cantor and Hodge's selection of essays [67] .

By the word "aether" or "ether" (we will equivalently interchange the two terms) one has always been indicating an elusive substance, different from ordinary bodies, that eventually fills the parts of space which are apparently empty. Invoked to explain the transmission of forces and the propagation of light, represented in different forms in different epochs, this entity has strongly influenced the modern philosophical and scientific

thought, from Descartes onward. The origin of the term, however, has its
roots in the antique Greek philosophy. In Aristotle's De Caelo, the aether
is introduced as the constitutive matter of the heavens. It is in a per-
petual state of circular motion and is characterized by its incorruptibility,
not being subject to aging, alterations and other affections which are typical of ordinary matter. For this reason, for Aristotle, the aether was the
first body: "Considering the first body as another substance, besides hearth,
fire, air and water, the ancients named the highest place aether (*aither*)
and gave this name because it always flows in the eternity of time (from
aei=always and *thein*=to flow)".

2.2   Descartes

Two thousands years later, Aristotle's circular aether became Descartes'
vortexes of *subtle matter* supporting the motion of the planets around the
sun. To explain how Descartes arrived to this view, let us first remind his
starting point: to consider extension as the sole essential property of matter
and matter as a necessary condition of extension. With this premise, for
him, the mere existence of bodies separated by a distance was a proof of the
existence of a continuous medium between them: "If someone asks what
would happen if God were to take away any single body contained in a
vessel, without allowing any other body to take the place of what had been
removed, the answer must be that the sides of the vessel would, in that
case, have to be in contact. For when there is nothing between two bodies
they must necessarily touch each other. And it is a manifest contradiction
for them to be apart, or to have a distance between them for every distance
is a mode of extension and therefore cannot exist without an extended
substance" [68].

  Thus, the impression of a vacuum is just an illusion. Descartes explains
it by first recalling the relative meaning of vacuum (vacuum of what?) and
then making examples which show how our senses are unable to perceive
some kinds of bodies and movements "...when we say that in a place there
is vacuum, it is evident that we do not mean there is nothing in that place
or space, but only that there is nothing of what we presume there should
be" [69].

  What there appears to be empty space should actually be thought as
filled by a form of *subtle matter* whose constituents are imperceptibly small
and in a very rapid agitation. These characteristics are similar to those of
a fluid: "I conceive the subtle matter as a continuous fluid which occupies

the space which is not occupied by other bodies" [70]. This aspect of the subtle matter is often a source of ambiguities, also for the differences which exist between the original version contained in The World or Treatise on the Light (1633) and the later version in the Principles of Philosophy (1647). A useful reference to deepen the subject is Lynes article [71].

An essential point, for Descartes, is the existence of only one form of matter. However, during the cosmogonic process, from this original matter there arose three fundamental elements, say E1, E2 and E3. Their constituents are characterized by different sizes and different speeds of motions. Subtle matter is of the kind E1, which forms the bright matter of the sun and of the stars and of the kind E2 which fills the interplanetary and interstellar space. Instead, E3 represents the dense matter of earth and planets. The partition in elements is such not to leave any empty space. Thus, the subtlest E1 particles go and occupy the interstices left by the E2 particles which occupy the space left over by the grosser E3 matter. This explains why "a vessel full of gold or lead does not contain more matter than when we think it is empty; and this may seem strange to many people whose reason does not reach much further than their fingers and who think that there is nothing in the world apart from what they touch" [72].

For Descartes, the subtle matter also became a tool to explain the propagation of light, of heat and of interactions among the observable forms of matter without introducing hidden influences through the *nothing*. For example, to describe gravity it is essential that the world is a *plenum* to give rise to the concept of *vortex*: "All places are full of matter and the same portion of matter always takes up the same amount of space. It follows from this that a body can move only in a closed loop of matter, a ring of bodies all moving together at the same time: a body entering a given place expels another, which moves on and expels a third body, and so on, until finally a body closes the loop by entering the place left by the first body at the precise moment when the first body is leaving it" [73].

This leads directly to the anticipated picture where the planets (E3 matter) were believed to float in a vortex of subtle matter E2 thus explaining why Kepler's orbit were all lying in the same plane, an unsolved aspect in Newton's theory. Today, the explanation is found in the Kant-Laplace theory of the original vortex from which the solar system was created. In this sense, "Kant resumed the valid part of the Cartesian theory by assuming the existence of the vortex and filling the unsolved gaps of the Newtonian doctrine. Thus, Descartes was a precursor of the present theory of the genesis of planets" [74].

2.3   Newton

"A Frenchman who arrives in London will find Philosophy, like everything else, very much changed there. He had left the world as a *plenum* and now he finds a *vacuum*". This fragment, taken from one of Voltaire's Philosophical Letters (1730), is reported by Whittaker [66] to describe the big difference between the Newtonian and Cartesian views just after the death of Newton. To appreciate the various aspects, it is also useful the other fragment: "It is the language used, and not the thing in itself, that irritates the human mind. If Newton had not used the word *attraction* in his admirable philosophy, everyone in our Academy would have opened his eyes to the light; but unfortunately he used in London a word to which an idea of ridicule was attached in Paris; and on that alone he was judged adversely, with a rashness which will some day be regarded as doing very little honour to his opponents".

This also explains the strong resistance to the Newtonian doctrine of universal gravitation. In spite of being the first theory explaining Kepler's laws as well as terrestrial gravity, it was considered the effect of an occult *action at a distance* and, as such, unacceptable for the disciples of Descartes.

However, one may ask whether this picture corresponds to Newton's true conception. To try to understand, let us first consider what he writes in his Principia (1687). There, Newton emphasizes that he only determined how the gravitational force depends on the relative positions of the bodies (the inverse-square law). Instead, to explain the process by which the gravitational action is effected is a quite distinct step which he did not attempt to make. Thus, about the cause of of gravity, Newton says: "I frame no hypotheses...To us it is enough that gravity does really exist, and act according to the laws which we have explained" [75].

Though, gravity must be produced by the action of some *agent*. On this point, in one of his letters to Bentley (1692-1693), Newton was writing: "That gravity should be innate, inherent, and essential to matter, so that one body can act upon another at a distance, through a vacuum, without the mediation of anything else, by and through which their action and force may be conveyed from one to another, is to me so great an absurdity, that I believe no man who has in philosophical matters a competent faculty of thinking can ever fall into it... Gravity must be caused by an agent acting constantly according to certain laws, but whether this agent be material or immaterial is a question I have left to the consideration of my readers".

This has been usually understood (see e.g. Maxwell [76]) to imply that Newton was not considering gravity as an action at a distance through the vacuum. Still, the mentioned material-immaterial ambiguity suggests, see Henry [77], that, for Newton, an action at a distance through a vacuum could still be possible if mediated by an *immaterial* agent. The fact that gravity is not innate, inherent and essential to matter could then just mean that gravity cannot be explained by the same physical properties, as hardness, extension, inertia, which are intrinsic to the nature of matter. For this reason, matter could not attract other matter at a distance without a *secondary* agent which directly derives from God (who is the primary cause)[1]. In this sense, "what Newton is trying to do is to make sure that the observed reality of action at a distance can be used to prove the existence of God" [77].

In this alternative interpretation, the immaterial agent of gravity would somehow resemble that subtle *electric Spirit* mentioned in the General Scholium at the end of the Principia (1713): "And now we might add something concerning a certain most subtle Spirit which pervades and lies hid in all gross bodies; by the force and action of which Spirit the particles of bodies mutually attract one another at near distances, and cohere, if contiguous; and electric bodies operate to greater distances, as well repelling as attracting the neighbouring corpuscles; and light is emitted, reflected, refracted, inflected, and heats bodies; and all sensation is excited, and the members of animal bodies move at the command of the will, namely, by the vibrations of this Spirit, mutually propagated along the solid filaments of the nerves, from the outward organs of sense to the brain, and from the brain into the muscles. But these are things that cannot be explained in few words, nor are we furnished with that sufficiency of experiments which is required to an accurate determination and demonstration of the laws by which this electric and elastic Spirit operates" [75]. Notice that here Newton is not using the Cartesian term *subtle matter* or the term *aether* and one is faced with the same material-immaterial ambiguity of the letter to Bentley.

On the other hand, to this alternative interpretation, one may easily object that, in some of the Queries at the end of Optics (1717), Newton

---

[1]Newton in the General Scholium brought God into natural philosophy, as the cause of the order in the world. For him the action of God was necessary to explain why the orbits of all planets and satellites lie in the same plane: "It is not to be conceived that mere mechanical causes could give birth to so many regular motions, since the comets range over all parts of the heavens in very eccentric orbit...This most beautiful system of the sun, planets and comets, could only proceed from the counsel and dominion of an intelligent and powerful Being" [75].

formulated a specific dynamical model where he was trying apparently to explain gravity by means of a material aether (not a spirit). Truly enough, this medium has some exceptional features and, for this reason, is very different from all other known media. To appreciate this aspect, let us first consider the 18th Query, where Newton argues about a medium which may enter in the transmission of heat and light: " If in two large tall cylindrical Vessels of Glass inverted, two little Thermometers be suspended so as not to touch the Vessels, and the Air be drawn out of one of these Vessels, and these Vessels thus prepared be carried out of a cold place into a warm one; the Thermometer *in vacuo* will grow warm as much, and almost as soon as the Thermometer which is not *in vacuo*. And when the Vessels are carried back into the cold place, the Thermometer *in vacuo* will grow cold almost as soon as the other Thermometer. Is not the Heat of the warm Room convey'd through the *Vacuum* by the Vibrations of a much subtler Medium than Air, which after the Air was drawn out remained in the Vacuum? And is not this Medium the same with that Medium by which Light is refracted and reflected, and by whose Vibrations Light communicates Heat to Bodies, and is put into Fits of easy Reflexion and easy Transmission? And do not the Vibrations of this Medium in hot Bodies contribute to the intenseness and duration of their Heat? And do not hot Bodies communicate their Heat to contiguous cold ones, by the Vibrations of this Medium propagated from them into the cold ones? And is not this Medium exceedingly more rare and subtile than the Air, and exceedingly more elastic and active? And doth it not readily pervade all Bodies? And is it not (by its elastic force) expanded through all the Heavens?" [78].

The same imperceptible medium could also account for gravity. To this end, there should be substantial changes in its density. At the same time, as compared to air, its elasticity should be much larger to explain why the speed of light is so much larger than the speed of sound. These aspects are treated in the 21st Query: "Is not this Medium much rarer within the dense Bodies of the Sun, Stars, Planets and Comets, than in the empty celestial Spaces between them? And in passing from them to great distances, doth it not grow denser and denser perpetually, and thereby cause the gravity of those great Bodies towards one another, and of their parts towards the Bodies; every Body endeavouring to go from the denser parts of the Medium towards the rarer? For if this Medium be rarer within the Sun's Body than at its Surface, and rarer there than at the hundredth part of an Inch from its Body, and rarer there than at the 50th part of an Inch from its Body, and rarer there than at the Orb of Saturn; I see no reason why the Increase of density should stop any where, and not rather be continued through all distances from the Sun to Saturn, and beyond. And though this Increase of density may at great distances be exceeding slow, yet if the elastic force of this Medium be exceeding great, it may suffice to impel Bodies from the denser parts of the Medium towards the rarer, with all that power which we call Gravity. And that the elastic force of this Medium is exceeding great, may be gather'd from the swiftness of its Vibrations. Sounds move about 1140 English Feet in a second Minute of Time, and in seven or eight Minutes of Time they move about one hundred English Miles. Light moves from the Sun to us in about seven or eight Minutes of Time, which distance is about 70000000 English Miles, supposing the horizontal Parallax of the Sun to be about 12". And the Vibrations or Pulses of this Medium, that they may cause the alternate Fits of easy Transmission and easy Reflexion, must be swifter than Light, and by consequence above 700000 times swifter than Sounds. And therefore the elastic force of this Medium, in proportion to its density, must be above 700000 x 700000 (that is, above 490000000000) times greater than the elastick force of the Air is in proportion to its density[2]".

But how could one explain this? And how could this be reconciled with the vanishing resistance that an aether should oppose to the motion of the planets? Newton argues that this could derive from two aspects, the extremely small size of the aether particles and the strong repulsion among them: "And so if any one should suppose that Aether (like our Air) may contain Particles which endeavour to recede from one another (for I do not know what this Aether is) and that its Particles are exceedingly smaller than those of Air, or even than those of Light: The exceeding smallness of its Particles may contribute to the greatness of the force by which those Particles may recede from one another, and thereby make that Medium exceedingly more rare and elastick than Air, and by consequence exceedingly less able to resist the motions of Projectiles, and exceedingly more able to press upon gross Bodies, by endeavouring to expand it self" [78].

The same concept is further developed in the 22nd Query: "May not Planets and Comets, and all gross Bodies, perform their Motions more freely, and with less resistance in this Aethereal Medium than in any Fluid, which fills all Space adequately without leaving any Pores, and by consequence is much denser than Quick-silver or Gold? And may not its resis-

---

[2]Here Newton is using the relation that, in an elastic medium, the squared velocity of waves is u^2 =elasticity/density.

tance be so small, as to be inconsiderable? For instance; If this Aether (for so I will call it) should be supposed 700000 times more elastick than our Air, and above 700000 times more rare; its resistance would be above 600000000 times less than that of Water. And so small a resistance would scarce make any sensible alteration in the Motions of the Planets in ten thousand Years" [78].

This basic idea of a gravitational aether made up of very small particles was not completely new, having already been proposed by Huygens in his Discourse on the Cause of Gravity (1690): "The extreme smallness of the parts of our fluid matter is still absolutely necessary to provide a reason for one notable aspect of gravity, namely that massive bodies, enclosed on all sides in a vessel of glass, metal, or any other material, are always found to have the same weight. So the matter that we have said is the cause of gravity must pass very easily through all bodies, even the most dense, with the same ease as through the air". Presumably, this had some influence on Newton, see his comment, in response to Leibniz (1693), about Huygens' treatise: "If any one should explain gravity and all its laws by the action of some subtle medium, and should show that the motions of the planets and comets were not disturbed by this matter, I should by no means oppose it" [79].

Thus, one might be tempted to conclude that, indeed, Newton had in mind a definite material model for the transmission of gravity by finally solving the material-immaterial ambiguity. But yet, another argument indicates that Newton's aether has not the obvious interpretation of an ordinary material medium. Again, as stressed by Heimann in his contribution [80] to Cantor and Hodge collection of essays [67], this derives from the basic distinction that Newton operates (in the 31st Query) between gravity, defined as an *active Principle*, and those passive principles related to the intrinsic properties of matter as inertia: "The Vis inertiae is a passive Principle by which Bodies persist in their Motion or Rest, receive Motion in proportion to the Force impressing it, and resist as much as they are resisted. By this Principle alone there never could have been any Motion in the World. Some other Principle was necessary for putting Bodies into Motion" [78].

On this basis, the necessary existence of the active Principles leads Newton to a cosmological view where "God in the Beginning form'd Matter in solid, massy, hard, impenetrable Particles" which are "incomparably harder than any porous Bodies compounded of them". These particles "have not only a Vis inertiae, accompanied with such passive Laws of Motion but also are moved by certain active Principles, such as is that of Gravity" [78] .

Thus, the gravitational aether, made up of these original particles, embodies the "active principle establishing the intelligibility of the phenomenon of gravity" and, in this sense, acquires a *theological function* [80].

For these reasons, Newton's aether can hardly be considered a simple material medium (as for Descartes and Huygens). For instance, in Westfall's interpretation [81] , Newton's aether represents an immaterial medium which "could move bodies without offering resistance to them in turn", a property which he formerly attributed to the omnipresence of God ("bodies find no resistance from the omnipresence of God" [75]). From this, Westfall deduces the ultimate identification of the aether with the "Sensorium of God", the concept Newton had introduced to indicate how God moves and controls the world in analogy to how we move and control our body.

We believe that, apart from Newton's theological views, this conception may also be due to the aether's exceptional features. Most notably, the possibility of producing the strong mechanical pull needed to keep the planets in their orbit while, at the same time, opposing a vanishing resistance to their motion. For Newton, who was not aware of the existence of superfluid media, this was an extraordinary challenge. Thus he remained always reluctant to definitely resolve the material-immaterial enigma. This persistent ambiguity of the discoverer of the universal gravitation would have exerted an important influence on the subsequent development of the scientific thought [80].

2.4   Kant

Kant's conception of ether is not generally known but deserves to be mentioned for at least two reasons. On the one hand, Kant affirms that the existence of the ether can be proven on a pure axiomatic basis. On the other hand, the ether itself is imperceptible so that the resulting picture has interesting analogies with the modern view of a physical, structured space which cannot be thought as trivially empty insofar it provides the essential support for the known, existing forces. For Kant, the forces are both the external forces, which can move the bodies, and the internal forces, through which matter limits itself in shape and size. In this sense, Kant's ether is well described as "a collectively moving material, continuously expanding and constantly agitating...a compositionally plastic, intrinsically structural substrate of dynamical force" [82].

The precise location of Kant's ether deduction is in his unfinished work, generally called as the "Opus Postumum", to which he worked during his last years until his death in 1804. The origin of this thought, however, dates back to several years before, namely to his "Metaphysical Foundations of Natural Science" in 1786 and to his "Critique of Pure Reason" in 1787.

To introduce the argument, let us first recall two essential aspects of Kant's doctrine, namely the relation between mathematics and natural science and the role of experiment. For Kant, a natural science can only be truly "scientific" to the extent that applies mathematics to his objects: "In any special doctrine of Nature there is only as much genuine science as there is mathematics" [83]. The explanation for this, however, lies outside of mathematics because "Natural science properly so-called presupposes the metaphysics of Nature, i.e. pure rational knowledge from mere concepts" [83]. It is illusory to think that one can do without: "So all natural philosophers who have wanted to proceed mathematically in their work have availed themselves (without realizing it) of metaphysical principles; they had to do so, despite their solemn declarations that metaphysics has no claims on their science" [83].

Therefore metaphysics, which by definition has no empirical content, "might deal with the laws that make possible the concept of a nature in general, without bringing in any specific object of experience, and therefore not saying anything specific about any particular kinds of empirical object" [83]. This fundamental role of metaphysics gives an autonomy to the theory of knowledge, with respect to the experimental results, which Kant summarizes by saying: "Reason, in order to be taught by nature, must approach nature with its principles in one hand, according to which the agreement among appearances can count as laws, and, in the other hand, the experiment thought out in accord with these principles, in order to be instructed by nature not like a pupil, who has recited to him whatever the teacher wants to say, but like an appointed judge who compels witnesses to answer the questions he puts to them" [84].

Though, on this basis alone, Kant had to admit that pure reason can only describe those properties which belong to a nature in general (such as the regularity of phenomena in space and time). Instead, about definite laws, those which concern phenomena empirically determined, there can be no knowledge without experience. At the same time, concluding that physics obtains these laws solely from experience would mean to accept the idea of pure "empirical laws", and thus fall under Hume's critics (how can any collection of particular experiences acquire a universal value?). Then, what to do? To stop giving a rigorous character to the knowledge of nature? This was Kant's attitude at the time of the Critique. Instead, later on, in the Opus Postumum, he tried to conceive a "Transition (*Übergang*) from metaphysical principles of natural science to physics".

At a first sight, this idea may be considered (by Kant himself) as "absurd" or "senseless". In fact, on the basis of our intellect alone, we should anticipate some elements which can only be confirmed *a posteriori*, i.e. by our direct experience of phenomena. However, Kant insists, this is unavoidable if we want to give full scientific value to what, otherwise, would represent a mere compilation, mere empiricism. In his view, the *Übergang* must not consist of *a priori* concepts alone, otherwise it would be pure metaphysics, but cannot be based on pure empiric representations either. Instead, it must concern the general system of the forces which are responsible for the motion of matter and, as such, are at the base of our experience. This is the context where to frame Kant's ether which represents the crucial element for the *Übergang*.

To this end, Kant first defines physics as the "doctrinal system of empirical knowledge in general", i.e. which "has as its object things whose cognition is only possible through experience" [85]. Now, the basic empirical apperceptions are "a reaction to movable in space (matter), insofar as a sensible external object, and to its motion", so that the existence of moving forces, i.e. that are able to move the bodies in space, and thus are able to impress the senses, is the foundation of the apperception of external objects. For this reason, physics is the "doctrinal system of the moving forces of matter" [85] . These forces have their ultimate origin in an all-pervading ether and, for this reason, should be thought as *local* forces, i.e. that act on a material particle at the given place of its location. Notice the analogy with the modern view of a *force field*.

Truly enough, in spite of these definite premises, it is not easy to summarize Kant's reflections on the ether. The Opus Postumum has no systematic order and there are many repetitions of the same concepts in slightly different terms. Thus, to represent Kant's ether view, we have chosen the following citations [85]:

"The basis of all possible perceptions of the moving forces of matter in space and time is the concept of an elementary material, distributed everywhere in cosmic space, attracting and repelling only in its own parts, and which is continuously internally self moving. Its concept is made into the sole principle (i.e. the principle of motion) for the possibility of experience".

"This form of a universally distributed, all-penetrating world material, which is in continuous motions in its own location, characterizes the original moving matter as a real existing material according to the principle of the possibility of the experience itself".

"There exists a matter, distributed in the whole universe as a continuum, uniformly penetrating all bodies and filling all spaces, be it called ether, caloric or whatever, which is not a hypothetical material (for the purpose of explaining certain phenomena, and more or less conjuring up causes for given effects); rather it can be recognized and postulated *a priori* as an element necessarily belonging to the transition from the metaphysical foundation of natural science to physics".

For Kant, the ether is thus a logically necessary entity which is at the base of the fundamental forces and, as such, makes possible our experience. At the same time, it cannot be directly perceived. The ether, in fact, is at the base of all our perceptions and no sense can perceive itself. For this reason, Kant's ether represents an intermediate concept between matter and space. We could say a form of *physical space*, i.e. containing those known forces which are necessary for our experience ("hypostatized space").

2.5   Young and Fresnel

Although, apparently, advocating in the "Optical Queries" the idea that light was originating from the vibrations of an underlying aether, Newton remained unable to put this view in a coherent form. One of the main reasons was, for him, the inability of a wave theory to explain the rectilinear propagation of rays which instead was immediate in the corpuscular theory. This was for him a problem, in spite of Huygens' discovery that two rays of light can cross each other without any reciprocal disturbance, as two trains of water waves do, at variance from what happens with two bunches of corpuscles where disturbances of some kind would necessarily occur. Huygens's explanation was based on that Principle, now called after his name, that every point on which light impinges is to be regarded as the source of a new spherical wave of light. However, Newton's great authority was an obstacle to the full acceptance of Huygens construction. Only at the beginning of 19th century the wave theory finally became the accepted picture. This was the consequence, mainly, of the work of two brilliant scientists: Thomas Young and Augustine Fresnel.

Young's initial observation [86] was that the remarkable constancy of the velocity of light was difficult to understand in the corpuscular theory but was extremely natural in the wave theory, because all disturbances are transmitted through an elastic fluid with the same velocity. The thorough exposition of Young's theory is given in his 1802 Bakerian Lecture [87].

Young starts by saying that from the reading of Newton's papers, he was led to the view that Newton had abandoned the corpuscular or emission-theory of light for what is nearly the undulatory theory: "For this reason, after having briefly stated each particular position of my theory, I shall collect, from Newton's various writings, such passages as seem to be the most favourable to its admission; and, although I shall quote some papers which may be thought to have been partly retracted at the publication of the optics, yet I shall borrow nothing from them that can be supposed to militate against his maturer judgement" [87]. This study of Newton's papers "converted the prepossession which I [Young] before entertained for the undulatory system of light into a very strong conviction of his truth and sufficiency, a conviction which has been since most strikingly confirmed by an analysis of the colours of striated substances" [87].

After this premise, Young formulates his famous four hypotheses [87]:

HYPOTHESIS I. A luminiferous Ether pervades the Universe, rare and elastic in a high degree.

HYPOTHESIS II. Undulations are excited in this Ether whenever a Body becomes luminous.

HYPOTHESIS III. The Sensation of different Colours depends on the different frequency of Vibrations, excited by Light in the Retina.

HYPOTHESIS IV. All material Bodies have an Attraction for the ethereal Medium, by means of which it is accumulated within their Substance, and for a small Distance around them, in a State of greater Density, but not of greater Elasticity.

From these hypotheses, he arrives to a coherent system (expounded in the set of PROPOSITIONS I-VIII and COROLLARIES I-V) which culminates in his final

PROPOSITION IX. Radiant Light consists in Undulations of the luminiferous Ether.

With the following remark: "This proposition is the general conclusion from all the preceding; and it is conceived that they conspire to prove it in as satisfactory a manner as can possibly be expected from the nature of the subject. It is clearly granted by NEWTON, that there are undulations, yet he denies that they constitute light; but it is shown in the three first Corollaries of the last Proposition, that all cases of the increase or diminution of light are referable to an increase or diminution of such undulations, and that all the affections to which the undulations would be liable, are distinctly visible in the phenomena of light; it may therefore be very logically inferred, that the undulations are light" [87]. This basis of Young's system, with his description of interference phenomena, represents a milestone in the history of science.

Though, also Young's formulation was not immediately accepted. In fact, diffraction was not satisfactory explained, there was no explanation for the phenomenon of *polarization*, double refraction seemed to require two different kinds of ether [66]. These issues were the main object of Fresnel's research which was decisive to establish the wave conception of light.

Fresnel's first investigation concerned the study of diffraction [88] by which he hoped to get a positive support for the undulatory view of light. The key observation depended on the use of the principle of interference. In fact, Fresnel understood that, if light were really undulatory, bright zones would occur where the vibrations constituting light are in phase while zone of darkness would correspond to places where the vibrations are out of phase and cancel each other. To put such concepts in a mathematical form, Fresnel derived formulas, correlating the positions of the bands to the light wavelength and the difference of optical path, which were confirmed by experiments.

Meanwhile, both Young and Fresnel had become aware of Malus' discovery that light, when refracted at the surface of transparent media, possesses in some degrees a directional property called, by Malus, *polarization*. Together with Arago, Fresnel original experiments tried to detect observable differences in the interference patterns between beams of polarized light. Initially, experiments showed no difference with ordinary light. However, more refined experiments showed that in circumstances where ordinary light would interfere, rays polarized in mutually perpendicular planes have no effect on one another.

The explanation proposed by Young consisted in resuscitating a model considered by (the young) Bernoulli [66] more than eighty years before: light vibrations are *transverse*, i.e. the oscillations of ether particles are perpendicular to the direction of wave propagation. Young's idea was first contained in his letter to Arago of 1817 and then more thoroughly exposed in an article for the supplement of the Encyclopedia Britannica of the same year: "If we assume as a mathematical postulate, on the undulating theory, without attempting to demonstrate its physical foundation, that a transverse motion may be propagated in a direct line, we may derive from this assumption a tolerable illustration of the subdivision of polarized light by reflection in an oblique plane, by supposing the the polar motion to be resolved into two constituents which fare differently at reflection" [66].

In a further letter to Arago of 1818 Young compared the transverse vibrations of light to the undulations of a cord agitated by one of its extremities. This letter was shown by Arago to Fresnel who at once realized this was the true explanation for the non-interference of beams polarized in perpendicular planes [66]. Indeed, if the vibrations of each beam are resolved into three components, one along the ray and the other two at right angles to it is obvious, from the Arago-Fresnel experiment, that the components in the direction of the ray must vanish.

Whittaker observes that, at that time, the formal theory of wave propagation in elastic media was yet unknown. Therefore, the undulatory theory was still based on the simple model of longitudinal oscillations, as for sound waves. As such, the idea of transverse waves had no physical justification. However, Fresnel anticipated these formal developments. In fact, as a possible explanation, he observed that "the geometers who have discussed the vibrations of elastic fluids hitherto have taken into account of no accelerating forces except those arising from the differences of condensation or dilatations between consecutive layers" [88]. Instead, if the ether possesses some rigidity, i.e. the power of resisting to distortions, such as it is manifested by solid bodies it will be capable of transverse vibrations. The absence of longitudinal waves would then indicate only that the forces opposing condensation and rarefaction are much more powerful than those which oppose to distortion. In this way, Fresnel arrived to conceive the ether as a nearly incompressible medium where the velocity of propagation of longitudinal waves is so much larger than the velocity of light that a practical constant density is maintained everywhere.

The other fundamental Fresnel's contribution is associated with the so called *Fresnel's drag*. The starting point of this study was Arago's result that starlight is refracted by a prism by the same amount as is light from a terrestrial source, regardless of the orientation of the prism to the direction of the earth in its orbit. This result was very difficult to understand on the basis of the standard explanation of stellar aberration discovered by Bradley in 1728. According to this standard view, the periodic apparent motion of a star was due to the difference of the velocity of light entering a telescope in different observations performed along the earth orbit. Therefore, Arago expected the velocity of light relative to the prism to change and, on the basis of "Newton's principle", the angle of refraction to change as well [89]. But he found that the earth motion was not producing any change in the refraction of rays originating from a given star.

To explain this result, Fresnel adopted Young's suggestion that the refractive power of transparent bodies depend on the concentration of the ether within them [66]. To make this more precise, he further assumed that the ether density in a body is proportional to the square of the refractive index $\mathcal{N}$. Thus if c denotes the velocity of light *in vacuo* (i.e. in free ether) and $c_1 = c/\mathcal{N}$ its velocity in the body (relatively to the system where the body is at rest) then the density ρ of ether in free space and the density ρ_1 in the body are related by $\rho_1 = \rho\mathcal{N}^2$. In this way, the density of ether carried along by the moving body is $\rho_1 - \rho = (\mathcal{N}^2 - 1)\rho$ while a quantity of ether ρ remains at rest. Therefore, if the body moves with a velocity v, the center of mass of the dragged ether moves with velocity $c_{\mathrm{drag}} = fv$, with $f = (\mathcal{N}^2 - 1)/\mathcal{N}^2$, and the velocity of light in the moving body (with respect to the free ether) is $c_1 + c_{\mathrm{drag}}$. As a consequence of these assumptions, if this expression is used to calculate the angle of refraction of starlight along the earth orbit [66], no change is observed during the year.

Fresnel's formula was experimentally confirmed by Fizeau in 1851 and by Hoek in 1868 with a slight different apparatus. The basis of these experiments was to study light propagation in water which was moving with velocity v_{lab} with respect to the laboratory. By splitting light in two beams (one along the direction of flowing water and one in opposite direction) and studying the interference of the two beams, Fizeau concluded that the two beams where propagating with velocities $c_1 \pm f v_{\mathrm{lab}}$ to very good accuracy (the sign depending on the direction of propagation).

Fresnel also deduced that, with his formula, there should be no change in the aberration angle if the telescope were filled with water — a result later confirmed by Airy in 1871. The consequence of all these developments was that the effects of the earth motion in the ether were *compensated* [89] by the Fresnel drag. Namely, when combined, all effects added up so as to produce a null effect to first order in the earth velocity.

2.6   Maxwell

To better appreciate the spirit of XIX century physics, let us imagine to attend a conference by James Clerk Maxwell at the Royal Society [76]: "I have no new discovery to bring before you this evening. I must ask you to go over very old ground, and to turn your attention to a question which has been raised again and again ever since men began to think. The question is that of the transmission of force. We see that two bodies at a distance from each other exert a mutual influence on each other's motion. Does this mutual action depend on the existence of some third thing, some medium of communication, occupying the space between the bodies, or do the bodies act on each other immediately, without the intervention of anything else?"

To introduce the matter, Maxwell first observes that from our ordinary experience "the action between bodies at a distance may be accounted for by a series of actions between each successive pair of a series of bodies which occupy the intermediate space". Therefore, even in those cases in which we cannot perceive the intermediate agency, it might more sensible to admit the existence of a medium which we cannot at present perceive rather than asserting that a body can act directly at a place where it is not.

To this remark, the advocates of the doctrine of action at a distance, reply by saying: "What right have we to assert that a body cannot act where it is not? Do we not see an instance of action at a distance in the case of a magnet, which acts on another magnet not only at a distance, but with the most complete indifference to the nature of the matter which occupies the intervening space? If the action depends on something occupying the space between the two magnets, it cannot surely be a matter of indifference whether this space is filled with air or not, or whether wood, glass, or copper, be placed between the magnets. Besides this, Newton's law of gravitation, which every astronomical observation only tends to establish more firmly, asserts not only that the heavenly bodies act on one another across immense intervals of space, but that two portions of matter, the one buried a thousand miles deep in the interior of the earth, and the other a hundred thousand miles deep in the body of the sun, act on one another with precisely the same force as if the strata beneath which each is buried had been non-existent. If any medium takes part in transmitting this action, it must surely make some difference whether the space between the bodies contains nothing but this medium, or whether it is occupied by the dense matter of the earth or of the sun".

To these arguments Maxwell replies, with his interpretation of Newton's thought, by recalling that, "with that wise moderation which is characteristic of all his speculations", Newton only determined how gravity depends on the relative positions of the heavenly bodies. To explain the process by which the gravitational action is effected was a quite distinct step which, in his Principia, he did not attempt to make. Instead, on the basis of Newton's letters to Bentley and Boyle and of the Queries at the end of Optics, one discovers that he attempted to explain gravitation by the pressure of a medium. However, being unable to obtain a satisfactory description of this medium he did not publish these investigations. For this reason, Maxwell concludes that "The doctrine of direct action at a distance cannot claim for its author the discoverer of universal gravitation".

Then, as far as magnetism is concerned, Maxwell turns to the method of investigation adopted by Faraday: "He with his penetrating intellect, his devotion to science, and his opportunities for experiments, was debarred from following the course of thought which had led to the achievements of the French philosophers, and was obliged to explain the phenomena to himself by means of a symbolism which he could understand". This new symbolism consisted in visualizing magnetic lines of force with the help of iron filings. With this method, these lines, extending in every direction from electrified and magnetic bodies, became in Faraday's mind physically distinct from the solid bodies from which they emanated. Thus, the experimenter "may trace the varying direction of the lines of force and determine the relative polarity, may observe in which direction the power is increasing or diminishing". Their observation shows that these filings form a system where the number of lines which pass through an area indicates the intensity of the force acting through the area. Also, "each individual line has a continuous existence in space and time. When a piece of steel becomes a magnet, or when an electric current begins to flow, the lines of force do not start into existence each in its own place, but as the strength increases new lines are developed within the magnet or current, and gradually grow outwards, so that the whole system expands from within".

By means of this new symbolism, Faraday defined with mathematical precision the whole theory of electromagnetism, in language free from mathematical technicalities. The motion which the magnetic or electric force tends to produce is invariably such as to shorten the lines of force and to allow them to spread out laterally from each other giving the idea of an underlying medium in a state of stress, "a tension, like that of a rope, in the direction of the lines of force, combined with a pressure in all directions at right angles to them. This is quite a new conception of action at a distance, reducing it to a phenomenon of the same kind as that action at a distance which is exerted by means of the tension of ropes and the pressure of rods". This is similar to what happens with the muscles of our bodies which tend to shorten themselves and at the same time to expand laterally. For similar reasons, Maxwell concludes, "we may regard Faraday's conception of a state of stress in the electromagnetic field as a method of explaining action at a distance by means of the continuous transmission of force, even though we do not know how the state of stress is produced".

To explore the properties of this medium which underlies electromagnetic phenomena, Maxwell observes that "one of Faraday's most pregnant discoveries, that of the magnetic rotation of polarized light, enables us to proceed a step farther". This may be described as follows: "Of two circularly polarized rays of light, precisely similar in configuration, but rotating in opposite directions, that ray is propagated with the greater velocity which rotates in the same direction as the electricity of the magnetizing current. It follows from this, as Sir W. Thomson has shown by strict dynamical reasoning, that the medium when under the action of magnetic force must be in a state of rotation — that is to say, that small portions of the medium, which we may call molecular vortices, are rotating, each on its own axis, the direction of this axis being that of the magnetic force". This explains the tendency of the magnetic lines of force to spread out laterally and to shorten themselves. It arises from the centrifugal force of the molecular vortices produced in the "electromagnetic medium".

Maxwell admits that "No theory of the constitution of the aether has yet been invented which will account for such a system of molecular vortices being maintained for an indefinite time without their energy being gradually dissipated into that irregular agitation of the medium which, in ordinary media, is called heat". However his conclusion is sharp: "Whatever difficulties we may have in forming a consistent idea of the constitution of the aether, there can be no doubt that the interplanetary and interstellar spaces are not empty, but are occupied by a material substance or body, which is certainly the largest, and probably the most uniform body of which we have any knowledge. Whether this vast homogeneous expanse of isotropic matter is fitted not only to be a medium of physical interaction between distant bodies, and to fulfil other physical functions of which, perhaps, we have as yet no conception, but also, as the authors of the Unseen Universe[3]

---

[3] Here Maxwell is referring to the book by B. Stewart and P. G. Tait, "The UNSEEN UNIVERSE: Physical Speculations on a Future State", MACMILLAN and Co. 1877.

seem to suggest, to constitute the material organism of beings exercising functions of life and mind as high or higher than ours are at present, is a question far transcending the limits of physical speculation".

2.7   The turbulent-ether model

Whittaker observes that Maxwell's and Thomson's electromagnetic ether was made up of small portions in rotatory motion and, as such, had already a first precursor in the model of Jean Bernoulli: a fluid ether containing an immense number of very small whirlpools. In this representation, usually denoted as the "vortex-sponge" model [66], the energy stored in the turbulent motion of the fluid becomes a source of elasticity. Thus the fluid behaves as a solid and can support the propagation of transverse waves whose speed c is comparable to the average speed of the internal vortical motion. In this sense, this type of medium, being capable to account mechanically for Maxwell's equations, was providing a model to conceive their Lorentz symmetry as an *emergent phenomenon*, i.e. emerging from a microscopic fluid whose elementary constituents are governed by Newtonian dynamics.

The turbulent-ether model has been more recently re-formulated by Troshkin [90] by adopting a modern formalism (see also [91] and [92]) and will be briefly summarized in the following. In this hydrodynamic analysis, one can start from the Euler equation that governs the velocity $\mathbf{u}$ of an incompressible fluid

$$\partial_t \mathbf{u} + (\mathbf{u} \cdot \nabla)\mathbf{u} + \nabla p = 0 \qquad (2.1)$$

where one assumes $\nabla \cdot \mathbf{u} = 0$ and p denotes the specific pressure. By introducing the Reynolds average of any given quantity $F(\mathbf{r}, t)$

$$\overline{F}(\mathbf{r}, t) = \frac{1}{T} \int_{t-T/2}^{t+T/2} F(\mathbf{r}, \tau) d\tau \qquad (2.2)$$

one can further separate out the fluctuating part $F'(\mathbf{r}, t)$

$$F(\mathbf{r}, t) = \overline{F}(\mathbf{r}, t) + F'(\mathbf{r}, t) \qquad (2.3)$$

such that $\overline{F'} = 0$. This type of averaging is performed over a time scale T which is much longer than the characteristic periods of the fluctuating components but also substantially shorter than the time scale of the evolution of the mean properties of the unsteady flow. In this way, one can first subtract from (2.1) its Reynolds-average (in component form)

$$\partial_t \overline{u}_i + \overline{u}_k \partial_k \overline{u}_i + \partial_i \overline{p} + \partial_k \overline{u'_i u'_k} = 0 \qquad (2.4)$$

then multiply the result by u'_j and finally write down an analogous equation by interchanging i and j. When this is added to the previous equation and the result is further averaged, one finds

$$\partial_t r_{ij} + \overline{u}_k \partial_k r_{ij} + r_{ik} \partial_k \overline{u}_j + r_{jk} \partial_k \overline{u}_i + h_{ij} = 0. \tag{2.5}$$

Here we have defined the Reynolds tensor r_{ij} (that vanishes in a laminar regime)

$$r_{ij} \equiv \overline{u'_i u'_j} \tag{2.6}$$

and the other auxiliary quantity

$$h_{ij} = \overline{u'_i \partial_j p'} + \overline{u'_j \partial_i p'} + \partial_k \overline{u'_i u'_j u'_k}. \tag{2.7}$$

At this stage, by following Troshkin [90], one truncates the system by first replacing

$$\overline{u'_i u'_j u'_k} \sim \overline{u'_i}\ \overline{u'_j u'_k} + \overline{u'_j}\ \overline{u'_i u'_k} + \overline{u'_k}\ \overline{u'_i u'_j} = 0 \tag{2.8}$$

and then assuming the sum of the first two terms in h_{ij} to be proportional to the deviation of the Reynolds tensor from its isotropic[4] equilibrium value $r_{ij}^{(0)}$

$$r_{ij}^{(0)} \equiv c^2 \delta_{ij} \tag{2.9}$$

namely

$$h_{ij} = \alpha(r_{ij} - r_{ij}^{(0)}) \equiv \alpha\, (\delta r)_{ij} \tag{2.10}$$

α being a proportionality constant. The origin of this closure scheme is the so called Rotta approximation which guarantees linear return to isotropy for decaying anisotropic turbulence [92–94].

Therefore, by assuming for all average quantities equilibrium values that solve exactly Eqs.(2.4) and (2.5) (with a vanishing equilibrium value of the average velocity) one can linearize in the small deviations from equilibrium $(\delta \overline{u})_i$, $(\delta r)_{ij}$ and $\delta \overline{p}$. Then Eqs.(2.4) and (2.5) take the form

$$\partial_t (\delta \overline{u})_i + \partial_i (\delta \overline{p}) + \partial_k (\delta r)_{ik} = 0 \tag{2.11}$$

and

$$\partial_t (\delta r)_{ij} + c^2 [\partial_i (\delta \overline{u})_j + \partial_j (\delta \overline{u})_i] + \alpha (\delta r)_{ij} = 0 \tag{2.12}$$

with the condition $\partial_i (\delta \overline{u})_i = 0$. These equations can be re-formulated in terms of some quantities $A_i \sim c(\delta \overline{u})_i$, $\phi \sim \delta \overline{p}$, $E_i \sim \partial_k (\delta r)_{ik}$ and $B_i =$

---

[4]This assumption reflects Kolmogorov's hypothesis [44] of local isotropy for fully developed turbulence. Namely, at small scales turbulence looses memory of any large-scale motion.

$(\nabla \mathrm{x} \mathbf{A})_i$ that express the deviations of the various average quantities from their equilibrium values and that, up to a proportionality factor, play the role of electromagnetic potentials and electromagnetic fields. In fact, from Eq.(2.11) and its curl one gets the first pair of Maxwell equations. On the other hand, by taking the divergence of Eq.(2.11), by differentiating Eq.(2.12) with respect to x_j (and summing over j) and using the relation $\Delta A_i = -(\nabla \mathrm{x} \nabla \mathrm{x} \mathbf{A})_i$ one gets the second pair with a charge density $\rho \sim -\Delta(\delta \bar{p})$ and a current density $J_i \sim \alpha E_i$. Thus, depending on the value of the parameter α, this type of 'vacuum' could behave as a conductor or as an insulator.

The important point is that, as in the original vortex-sponge model, the speed c of the transverse waves is proportional to the equilibrium isotropic speed of the turbulent motion. This means that, on a coarse-grained scale, the fluid is starting to behave as a solid (think of jets of water of sufficient speed). In this sense, the phenomenon of turbulence provides a conceptual transition from fluid dynamics to a different realm of physics, that of elasticity. At a more formal level, this is supported by the equivalence [95, 96] (velocity potential vs. displacement, velocity vs. distortion, vorticity vs. density of dislocations,...) that can be established between various systems of dislocations in an elastic solid and corresponding vortex fields in a liquid.

Chapter 3

The data published by Michelson and Morley, instead of giving a null result, show distinct evidence for an effect of the kind to be expected.

W. HICKS, Philosophical Magazine, 1902.

3.1 The idea of the ether drift

In his article for the Ninth Edition of the Encyclopedia Britannica [2], Maxwell also addressed the motion in the ether of dense bodies such as the earth. This part is essential because it gave the motivation for those *ether-drift* experiments which are the main subject of our study.

After having discussed why the medium which supports light propagation cannot be identified with the air itself or with the other solid transparent media (such as glass or crystals), Maxwell arrives to the main problem, namely trying to understand: " Whether when dense bodies are in motion through the great ocean of aether, they carry along with them the aether they contain, or whether the aether passes through them as the water of the sea passes through the meshes of a net when it is towed along by a boat". Thus, to obtain an experimental answer, one could argue that: " If it were possible to determine the velocity of light by observing the time it takes to travel between one station and another on the earth surface, we might, by comparing the observed velocities in opposite directions, determine the velocity of the aether with respect to these terrestrial stations. All methods, however, by which it is practicable to determine the velocity of light from terrestrial experiments depend on the measurement of the time required for the double journey from one station to the other and back, again, and the increase of this time on account of a relative velocity of the aether equal to that of the earth in its orbit would be only about one hundred millionth part of the whole time of transmission, and would therefore be quite insensible".

Maxwell does not report the details of his calculation. This, however, can be easily reconstructed as follows. Let us first consider the time t_1 for light propagating back and forth along a distance D in the direction of the earth motion. Assuming an earth velocity v with respect to the ether, according to classical physics light velocities are $c + v$ and $c - v$ respectively so that

$$t_1 = \frac{D}{c+v} + \frac{D}{c-v} = \frac{2D}{c} \frac{1}{1 - \beta^2} \tag{3.1}$$

where $\beta = v/c$. On the other hand, if the same path length D were in the perpendicular direction the corresponding time would now be[1] $t_2 = 2D/c$. Therefore, by estimating the velocity of the earth in the ether to be comparable to its orbital velocity $v \sim 30$ km/s, and expanding in powers of $\beta^2 = v^2/c^2$, one finds a relative difference

$$\frac{t_1 - t_2}{t_2} \sim \frac{v^2}{c^2} \sim 10^{-8}. \tag{3.2}$$

This effect was considered by Maxwell too small to be observed. However, later on, Maxwell's idea was taken seriously by Michelson and became the great challenge of his research.

3.2   Albert A. Michelson and his first 1881 experiment

Albert Abraham Michelson was born in Strzelno (now Poland at that time Prussia) on December 19, 1852 into a Polish-Jewish family. In 1855 the family emigrated to United States, first to New York City and then to San Francisco where he attended the high school. In 1869 he entered the U. S. Naval Academy where he graduated in 1873. After spending two

---

[1]Actually, as we shall see in a moment, the standard classical estimate of t_2 is different if one takes into account that a reflecting mirror, placed in the perpendicular direction, is also translating with velocity v. This gives a total difference $t_1 - t_2$ which is only one half of that computed above. In any case, Maxwell's order of magnitude 10^{-8} remains the same.

years on board of different ships, Michelson served as a physics teacher at Annapolis where he made his first determination of the speed of light with a demonstration for the students in November 1877. In 1879 he was transferred to the Nautical Almanac Office in Washington D. C. . There Simon Newcomb was the leading scientist who provided ample support for all Michelson's subsequent researches.

Robert Shankland [97] observes that, perhaps, the most important event occurring for Michelson while he was at the Nautical Almanac Office was his opportunity to study a letter dated 19 March 1879 from James Clerk Maxwell to David Peck Todd, then also a researcher associated with the same Office. The most relevant part of this letter was Maxwell's statement that "in the terrestrial methods of determining the velocity of light, the light comes back along the same path again, so that the velocity of the earth with respect to the ether would alter the time of the double passage by a quantity depending on the square of the ratio of the earth velocity to that of light, and this is quite too small to be observed" [97]. After several discussions with Todd and Newcomb, the possibility to detect the earth motion through the ether became the main goal at the base of his optical studies.

In 1880 Michelson left with his family for Europe and, after two brief stays in London and Paris, went on to the Institute of Physics at the University of Berlin where Hermann von Helmholtz was the director. The discussions with Helmholtz and the other colleagues, as well as the facilities of the laboratory, were certainly important ingredients for Michelson's fundamental intuition of detecting the earth motion through an optical interference method. However, interestingly, Swenson [98] argues that his experience as a sailor was probably another decisive element. Although there is little historical evidence to support this conclusion, still Michelson's "sea duty — an experience that inevitably invites the idea that one's ship is the world — made him better able to appreciate the possibility of measuring aether-drift in a manner analogous to the computation of true wind speed and direction" [98].

Swenson's idea is supported by Michelson's *incipit* of his 1881 paper [99]: "The undulatory theory of light assumes the existence of a medium called the ether, whose vibrations produce the phenomena of heat and light and which is supposed to fill all space. According to Fresnel, the ether, which is enclosed in optical media, partakes of the motion of these media, to an extent depending of their indices of refraction. For air this motion would be but a small fraction of that of the air itself and will be neglected. Assuming then that the ether is at rest, the earth moving through it, the time required for light to pass from one point to another on the earth surface would depend on the direction in which it travels". After this premise Michelson, shows how the tiny v^2/c^2 effect predicted by Maxwell could be measured with an apparatus which detects the interference pattern of two pencils of light propagating in orthogonal directions.

The necessary funds to realize the first experiment were provided by Alexander Graham Bell at the suggestion of Simon Newcomb and Michelson selected the firm of Schimdt and Haensch in Berlin to build the instrument (and the Maison Breguet in Paris for the optical flats [97]). This first apparatus is shown in Fig.3.1.

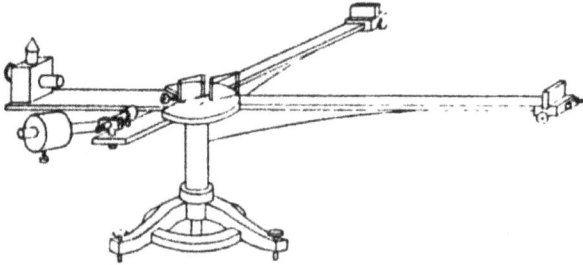

Fig. 3.1   *Michelson's 1881 original interferometer [99].*

The light beam from the source, at the extreme left of the figure, is divided by the beam splitter into two other rays that travel coherently at right angles and are then reflected by two mirrors and focalized in the telescope where produce an interference pattern.

As anticipated, the time t_1 required for light to cover a distance D along the drift direction, from the beam splitter to the mirror M_1 and return, is

$$t_1 = \frac{D}{c+v} + \frac{D}{c-v} = \frac{2D}{c}\frac{1}{1-\beta^2} \qquad (3.3)$$

To estimate the time t_2, required for light to travel from the beam splitter to the orthogonal mirror M_2 and return, requires instead to evaluate the displacement in the ether of M_2 which is produced by the earth motion. Since this total displacement along the earth motion is vt_2, by applying Pythagoras' theorem, the total path of light is

$$l_2 = 2\sqrt{(\frac{vt_2}{2})^2 + D^2} \qquad (3.4)$$

Therefore, by assuming that light has the same velocity c to make the to and fro journey, this gives a time

$$t_2 = \frac{l_2}{c}. \tag{3.5}$$

From this, by solving for t_2, one finally finds

$$t_2 = \frac{2D}{c} \frac{1}{\sqrt{1-\beta^2}} \tag{3.6}$$

which, for this orientation of the apparatus, gives a time difference Δt

$$\Delta t = t_1 - t_2 = \frac{2D}{c} \left[\frac{1}{1-\beta^2} - \frac{1}{\sqrt{1-\beta^2}} \right] \sim \frac{D}{c} \beta^2. \tag{3.7}$$

The corresponding time difference after a rotation of the apparatus by 90 degrees is instead

$$[\Delta t]_{\text{rot}} = [t_1 - t_2]_{\text{rot}} = \frac{2D}{c} \left[\frac{1}{\sqrt{1-\beta^2}} - \frac{1}{1-\beta^2} \right] \sim -\frac{D}{c} \beta^2. \tag{3.8}$$

Therefore, by transforming time differences in units of light wavelength $\Delta\lambda = c\Delta t$, a rotation of the apparatus by 90 degrees should produces a total *fringe shift* in the interference pattern

$$\left[\frac{\Delta\lambda(0)}{\lambda} - \frac{\Delta\lambda(\pi/2)}{\lambda} \right]_{\text{class}} = c \frac{\Delta t - [\Delta t]_{\text{rot}}}{\lambda} \sim \frac{2D}{\lambda} \beta^2. \tag{3.9}$$

For Michelson's original apparatus, the optical path was $D=120$ cm and in terms of wavelengths of white light ($\lambda \sim 5700$ Å) this distance is about $2 \cdot 10^6$. Therefore, for an earth speed of 30 km/s, the expected fringe shift for 90 degree rotations, with respect to the direction of the drift, was about 0.04.

For an easier comparison with the data it is useful, however, to give the general expression of the fringe pattern when the magnitude and direction of the drift, in the plane of the interferometer, are specified by a generic pair (v, θ_0) with, by convention, $\theta_0 = 0$ corresponding to North and $\theta_0 = \pi/2$ to East. In this case, as anticipated in Chapt.1, the classical prediction can be cast in the form

$$\left[\frac{\Delta\lambda(\theta)}{\lambda} \right]_{\text{class}} \sim \frac{D}{\lambda} \beta^2 \cos 2(\theta - \theta_0). \tag{3.10}$$

Thus, a 90 degree rotation from NS to EW gives

$$\left[\frac{\Delta\lambda(0)}{\lambda} - \frac{\Delta\lambda(\pi/2)}{\lambda} \right]_{\text{class}} \sim \frac{2D}{\lambda} \beta^2 \cos 2\theta_0 \equiv 2A_2^{\text{class}} \cos 2\theta_0 \tag{3.11}$$

while a 90 degree rotation from NE-SW to SE-NW gives

$$\left[\frac{\Delta\lambda(\pi/4)}{\lambda} - \frac{\Delta\lambda(3\pi/4)}{\lambda}\right]_{\text{class}} \sim \frac{2D}{\lambda}\beta^2 \sin 2\theta_0 \equiv 2A_2^{\text{class}} \sin 2\theta_0. \quad (3.12)$$

Also, one should keep in mind that, due to the definition of the two-way velocity of light, a genuine physical effect should produce a fringe pattern where

$$\frac{\Delta\lambda(\theta)}{\lambda} = \frac{\Delta\lambda(\theta+\pi)}{\lambda} \quad (3.13)$$

so that one should always average the data at θ and $\pi + \theta$ to extract the physical 2nd harmonic component. Finally, as anticipated in Chapt.1, by using these classical relations and normalizing to the expected amplitude for $v = 30$ km/s

$$A_2^{\text{class}} \sim \frac{D}{\lambda}\frac{(30 \text{ km/s})^2}{c^2} \sim 0.02 \quad (3.14)$$

one can extract from the experimental amplitude the *observable* velocity

$$v_{\text{obs}} \sim 30 \text{ km/s}\sqrt{\frac{A_2^{\text{EXP}}}{A_2^{\text{class}}}}. \quad (3.15)$$

Let us then look at Michelson's measurements. When the apparatus was placed on a stone pier in the Institute of Physics in Berlin, the first observations showed an extreme sensitivity of the instrument to the traffic vibrations so that measurements could not be carried on during the day [99]. To solve this problem, the experiment was first tried at night, with no much improvement, and then the apparatus was moved to the more quiet Astrophysics Observatory in Potsdam. Michelson arranged his apparatus in a windowless cellar of the basement of the Observatory where the measuring equipment was screened from vibrations and kept at a relatively constant temperature. In these new conditions, he was able to observe the fringes undisturbed and was himself astonished at the beauty of the interferometer performance [100].

The measurements were performed during April 1881. The observed average position of the fringes were: along North-South $\frac{\Delta\lambda(0)}{\lambda} = -0.003$, along North East-South West $\frac{\Delta\lambda(\pi/4)}{\lambda} = -0.007$, along East-West $\frac{\Delta\lambda(\pi/2)}{\lambda} = +0.001$, along North West-South East $\frac{\Delta\lambda(3\pi/4)}{\lambda} = +0.008$. Therefore the displacements by rotation of 90 degrees were

$$2A_2^{\text{EXP}} \cos 2\theta_0^{\text{EXP}} = -0.003 - (+0.001) = -0.004 \quad (3.16)$$

$$2A_2^{\text{EXP}} \sin 2\theta_0^{\text{EXP}} = -0.007 - (+0.008) = -0.015 \quad (3.17)$$

A fit to this set of data gives experimental values $A_2^{\mathrm{EXP}} \sim 0.0078$ and $\theta_0^{\mathrm{EXP}} \sim 127.5$ degrees (or $127.5 + 180$ degrees) with an observable velocity

$$v_{\mathrm{obs}} \sim 30 \text{ km/s} \sqrt{\frac{0.0078}{0.02}} \sim 19 \text{ km/s} \qquad (3.18)$$

which cannot be considered a negligible effect.

However, Michelson interpreted his measurements as typical instrumental effects. This is reported in his letter to Alexander Graham Bell of 17 April 1881: "The experiments concerning the relative motion of the earth with respect to the ether have just been brought to a successful termination. The result was however negative. At this season of the year the supposed motion of the solar system coincides approximately with the motion of the earth around the sun, so that the effect to be observed was at its maximum, and accordingly if the ether were at rest, the motion of the earth through it should produce a displacement of the interference fringes of at least one tenth the distance between the fringes; a quantity easily measurable. The actual displacement was about one one hundredth and this is assignable to the errors of the experiment. Thus the question is solved in the negative, showing that the ether in the vicinity of the earth is moving with the earth, a result in direct variance with the generally received theory of aberration" [98].

Analogous conclusions are expressed in his 1881 article: "The interpretation of these results is that there is no displacement of the interference bands. The result of the hypothesis of a stationary ether is thus shown to be incorrect, and the necessary conclusion follows that the hypothesis is erroneous. This conclusion directly contradicts the explanation of the phenomenon of aberration which has been hitherto generally accepted, and which presupposes that the earth moves through the ether, the latter remaining at rest" [99]. On this basis, Michelson concluded his paper by quoting a paragraph from a paper by Stokes about the possibility of an experiment that could decide between Fresnel's view of an immobile ether and his own theory based on a dragged-along ether. Reporting this quotation indicates that Michelson was convinced that he had provided such an experiment.

To partially explain Michelson's bold conclusions, we recall that, in this 1881 paper, he assumed that the earth velocity had no influence on the light beam traveling perpendicularly, i.e. he assumed $t_2 = 2D/c$. For this reason, as anticipated, his original expectation of the fringe shift under a 90 degree rotation was twice as large (0.08 rather than 0.04 for the pure orbital motion). In addition, he was composing the orbital earth motion with the solar motion, toward the constellation Hercules, and thus expecting an even larger average shift of about one tenth of a fringe. Finally, the data did not show the expected phase relationship on the basis of this presumed motion. Nevertheless, for the intrinsic uncertainty of these theoretical estimates, his conclusions seem too strong.

One more remark is about temperature effects. As reported in a letter that Michelson wrote to Newcomb at the end of 1880, Helmholtz had already warned him about this: "I had quite a long conversation with Dr. Helmholtz concerning my proposed method for finding the motion of the earth relative to the ether, and he said he could see no objection to it, except the difficulty of keeping a constant temperature" [98]. In fact, non-uniform temperature conditions can affect both the linear dimensions of the apparatus and the velocity of light along the optical paths due to asymmetric changes in the air density. Though, Michelson did not consider this as an unsurmountable problem. In fact, in his paper he was commenting as follows: "The principal difficulty which was to be feared in making these experiments, was that arising from changes of temperature of the two arms of the instrument. These being of brass whose coefficient of expansion is 0.00019 and having a length of about 1000 mm or 1.700.000 wave-lengths, if one arm should have a temperature only one-hundredth of a degree higher than the other, the fringes would thereby experience a displacement three times as great as that which would result from the rotation. On the other hand, since the changes of temperature are independent of the direction of the arms, if these changes were not too great their effect could be eliminated" [99]. Here Michelson is observing that temperature differences which are not synchronous with the rotation of the interferometer will largely be canceled in the averaging process of many measurements. On the other hand, a strictly periodic effect would survive. We will return to the intriguing role of temperature differences many times in the following.

3.3   The 1887 Michelson-Morley experiment

In spite of the strength of the conclusions of his 1881 paper, neither Michelson himself nor the scientific world generally ever considered the Potsdam trial as conclusive, although Lord Rayleigh, Lord Kelvin and Lorentz gave careful attention to this first ether-drift experiment [97]. This situation is summarized in a letter from Millikan to Shankland: "This first experiment must have been a very crude one, and it was only after Michelson got to

Case that he set up in connection with Morley the outfit which has since gone under the name of the Michelson-Morley Experiment" [97].

Michelson became professor of physics at the Case School in Cleveland in the fall of 1882. He found that Newcomb was still ready with funds and encouragement to support a more accurate determination of the velocity of light. Having obtained a grant of 1200$, Michelson worked constantly to improve his previous measurements in Annapolis and Washington. These efforts gave a result for the velocity of light (reduced to vacuum) of 299 850 km/s which remained the reference value for more than forty years (until the Michelson–Pease–Pearson 1927 determination) [97].

It was only in 1884 that he started seriously to think to improve his ether-drift experiment in connection with a series of lectures delivered by Lord Kelvin at the Johns Hopkins University in Baltimore. The lectures were attended by several physicists including Michelson and Edward W. Morley from the Western Reserve College of Cleveland. During the informal discussions of this period, Kelvin urged Michelson to improve on his Potsdam measurements by providing a new decisive test on the real magnitude of the ether-drift [97]. Also it is conceivable that, in this period, Morley (who at that time was a well recognized expert in chemistry and also in physics and mathematics) was drawn into the optical research problems.

However, the first joint research by Michelson and Morley was not on the repetition of the Potsdam experiment. It was instead dedicated to repeat Fizeau's experiment in water to precisely determine the influence of the motion of media in the propagation of light. In fact, the original Fizeau measurement was not conclusive to resolve some theoretical discrepancies in the actual amount of the drag.

The first Michelson and Morley joint experiment had this scope and gave results that were beautifully reproduced by the $1 - 1/n^2$ Fresnel formula to high accuracy. Therefore, the conclusion of their work was that "The result announced by Fizeau is essentially correct; and that *the luminiferous ether is entirely unaffected by the motion of the matter which it permeates*" [101].

With this preliminary experimental evidence, which was confirming the accepted theoretical scenario, Michelson and Morley were ready to start their repetition of the Potsdam experiment. To this end, an important motivation also came from Lord Rayleigh who urged Michelson to pay attention to a paper by Lorentz where, among the various things, the expected fringe shift had been recomputed and found to be only one-half of the value assumed in the 1881 paper. As we have seen in the previous section, a recalculation was giving an observable velocity of about 19 km/s so that an improved measurement was necessary. In his response, dated March 6 1887, Michelson acknowledged the importance of Lord Rayleigh's encouragement and promised to start at once with the new experiment.

The apparatus was improved in many respects to reduce the sensitivity to environmental disturbances and the distortions caused by rotations. To this end, the optical parts were mounted on a massive stone slab floating on mercury contained in a circular tank of cast iron, see Fig.3.2, placed in the basement of the Case University.

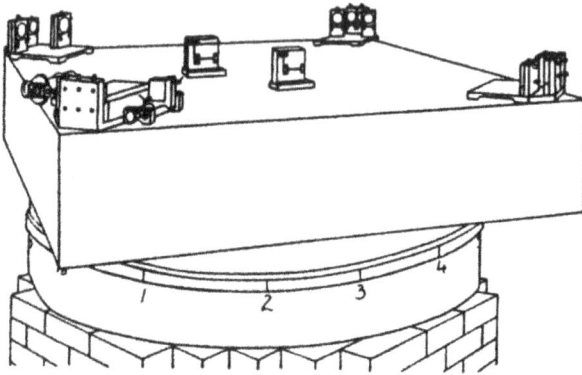

Fig. 3.2   *Michelson and Morley 1887 interferometer [1].*

With such support for floatation, by a gentle push of the observer's hand, the whole apparatus could be easily rotated slowly and continuously for hours at a time, a complete round being made in about 6 minutes (i.e. an angular speed of 1 degree/sec). The crucial part of the system was the beam splitter, or partially silvered mirror, which was dividing the source beam into two pencils of light traveling at right angles to each other. A micrometer screw was used to adjust one of the mirrors near the telescope to focus the fringes.

The optical paths were also increased by multiple reflections and reached a distance about ten times larger than the previous value (i.e. D=1100 cm vs. D=120 cm). In this way, by assuming an earth velocity $v \sim 30$ km/s and for a sodium lamp with characteristic yellow light with wavelength $\lambda = 5.89 \cdot 10^{-5}$ cm, the expected fringe shift by a 90 degree rotation about the drift direction was about 0.4 or, more precisely [4],

$$\left[\frac{\Delta\lambda(0)}{\lambda} - \frac{\Delta\lambda(\pi/2)}{\lambda}\right]_{\text{class}} \sim \frac{2D}{\lambda}\frac{v^2}{c^2} \sim 0.37 \qquad (3.19)$$

In early July 1887, Michelson and Morley performed their observations (July 8, 9 and 11 at noon and July 8, 9 and 12 at 6 P. M.), each observation consisting in 6 rounds of the interferometer for a total time of about 36 minutes. During the data taking, Michelson walked around the circuit and declared his reading of the fringe location at each of 16 equidistant points (i.e. one reading every 22.5 degrees) while Morley usually was sitting and recording the data [98]. The original data obtained during the six experimental sessions are reported in Fig.3.3.

NOON OBSERVATIONS.

	16.	1.	2.	3.	4.	5.	6.	7.	8.	9.	10.	11.	12.	13.	14.	15.	16*
July 8	44·7	44·0	43·5	39·7	35·2	34·7	34·3	32·5	28·2	26·2	23·8	23·2	20·3	18·7	17·5	16·8	13·7
July 9	57·4	57·3	58·2	59·2	58·7	60·2	60·8	62·0	61·5	63·3	65·8	67·3	69·7	70·7	73·0	70·2	72·2
July 11	27·3	23·5	22·0	19·3	19·2	19·3	18·7	18·8	16·2	14·3	13·3	12·8	13·3	12·3	10·2	7·3	6·5
Mean	43·1	41·6	41·2	39·4	37·7	38·1	37·9	37·8	35·3	34·6	34·3	34·4	34·4	33·9	33·6	31·4	30·8
Mean in w. l.	·862	·832	·824	·788	·754	·762	·758	·756	·706	·692	·686	·688	·688	·678	·672	·628	·616
	·706	·692	·686	·688	·688	·678	·672	·628	·616								
Final mean.	·784	·762	·755	·738	·721	·720	·715	·692	·661								

P. M. OBSERVATIONS.

	16.	1.	2.	3.	4.	5.	6.	7.	8.	9.	10.	11.	12.	13.	14.	15.	16*
July 8	61·2	63·3	63·3	68·2	67·7	69·3	70·3	69·8	69·0	71·3	71·3	70·5	71·2	71·2	70·5	72·5	75·7
July 9	26·0	26·0	28·2	29·2	31·5	32·0	31·3	31·7	33·0	35·8	36·5	37·3	38·8	41·0	42·7	43·7	44·0
July 12	66·8	66·5	66·0	64·3	62·2	61·0	61·3	59·7	58·2	55·7	53·7	54·7	55·0	58·2	58·5	57·0	56·0
Mean	51·3	51·9	52·5	53·9	53·8	54·1	54·3	53·7	53·4	54·3	53·8	54·2	55·0	56·8	57·2	57·7	58·6
Mean in w. l.	1·026	1·038	1·050	1·078	1·076	1·082	1·086	1·074	1·068	1·086	1·076	1·084	1·100	1·136	1·144	1·154	1·172
	1·068	1·086	1·076	1·084	1·100	1·136	1·144	1·154	1·172								
Final mean.	1·047	1·062	1·063	1·081	1·088	1·109	1·115	1·114	1·120								

Fig. 3.3 *Michelson-Morley original data as reported in [1] .*

The meaning of the various entries is the following. The first series of data (for both noon and evening observations) reports 17 readings corresponding to the various rotation angles $\theta = \frac{i-1}{16} 2\pi$ at steps of 22.5 degrees (the entry indicated by the first "16" corresponding to i=1 and thus to $\theta = 0$). These readings are first given in units of "divisions of the screwheads" [1]. Since "the width of the fringes varied from 40 to 60 divisions, the mean value being near 50" [1], it follows that "one division means 0.02 wavelength" [1]. For this reason, in the second series of entries, the average readings are transformed in units of wavelength after division by a factor 50.

Finally, the values obtained at each rotation angle θ are superimposed and then averaged with those at $\pi + \theta$ to construct a *second-harmonic* function, i.e. which is even under the exchange $\theta \to \pi + \theta$. This is necessary if the observed effects have to represent a measurement of the *two-way* velocity of light.

Now, as one can see, in their tables Michelson and Morley reported the readings with an accuracy of 1/10 of a division (example 44.7, 44.0, 43.5,..). This means that the nominal accuracy of the readings (averaged over the 6 turns) was ± 0.002 wavelengths. In fact, in units of wavelengths, they reported values such as 0.862, 0.832, 0.824... This estimate is confirmed by what is reported in Born's book [4]. There, when discussing the classically expected fractional fringe shift upon rotation of the apparatus by 90 degrees, about 0.37 as we have seen, Born says explicitly: "Michelson was certain that the one-hundredth part of this displacement would still be observable" (i.e. 0.0037). This would suggest a minimal detectable fringe shift of about ± 0.004. Apart from Born's quotation, this is confirmed by Holton [5] who reports an accuracy of "one part in ten billion" for the minimum light anisotropy detectable with the Michelson-Morley apparatus. Again, once multiplied by $(2D/\lambda) \sim 4 \cdot 10^7$, this is equivalent to minimal fringe shifts of ± 0.004. On this basis, this error magnitude (averaged over the 6 turns) will be assumed in the following[2].

By inspection of the data in Fig.3.3, one finds that all 17th entries say $E(17)$ (corresponding to the column under the second series of "16"), which were obtained after a complete rotation of the apparatus, are sizeably different from the starting entries, say $E(1)$ (indicated by the column under the first series of "16"). Therefore, if one would take these values to build a fringe-shift function $\Delta\lambda(\theta)$, the resulting $\Delta\lambda(\theta = 0)$ would be different from $\Delta\lambda(\theta = 2\pi)$. This difference was found in all ether-drift experiments and should be understood to obtain an unambiguous physical picture. So far, it has always been interpreted as a *thermal drift* due to changing temperature conditions in the laboratory which were affecting in a non-uniform way the solid parts of the apparatus[3]. Since, to good approximation, the observed

---

[2]To further confirm that such estimate should not be considered unrealistically small, we report Michelson's words from ref. [102]: "I must say that every beginner thinks himself lucky if he is able to observe a shift of 1/20 of a fringe. It should be mentioned however that with some practice shifts of 1/100 of a fringe can be measured, and that in very favorable cases even a shift of 1/1000 of a fringe may be observed".

[3]We emphasize that such thermal effects on the arms of the interferometer act on a longer time lag as compared to the other non uniformity which instead affects the density of the air along the optical paths [47].

Table 3.1 *The fringe shifts $\frac{\Delta\lambda(i)}{\lambda}$ for all noon (n.) and evening (e.) sessions of the Michelson-Morley experiment. The table is taken from ref. [38].*

i	July 8 (n.)	July 9 (n.)	July 11 (n.)	July 8 (e.)	July 9 (e.)	July 12 (e.)
1	−0.001	+0.018	+0.016	−0.016	+0.007	+0.036
2	+0.024	−0.004	−0.034	+0.008	−0.015	+0.044
3	+0.053	−0.004	−0.038	−0.010	+0.006	+0.047
4	+0.015	−0.003	−0.066	+0.070	+0.004	+0.027
5	−0.036	−0.031	−0.042	+0.041	+0.027	−0.002
6	−0.007	−0.020	−0.014	+0.055	+0.015	−0.012
7	+0.024	−0.025	+0.000	+0.057	−0.022	+0.007
8	+0.026	−0.021	+0.028	+0.029	−0.036	−0.011
9	−0.021	−0.049	+0.002	−0.005	−0.033	−0.028
10	−0.022	−0.032	−0.010	+0.023	+0.001	−0.064
11	−0.031	+0.001	−0.004	+0.005	−0.008	−0.091
12	−0.005	+0.012	+0.012	−0.030	−0.014	−0.057
13	−0.024	+0.041	+0.048	−0.034	−0.007	−0.038
14	−0.017	+0.042	+0.054	−0.052	+0.015	+0.040
15	−0.002	+0.070	+0.038	−0.084	+0.026	+0.059
16	+0.022	−0.005	+0.006	−0.062	+0.024	+0.043

magnitude of this drift was found to increase linearly with the rotation angle [7, 47], one can correct as described in Miller's paper [7]. Namely, by starting from each set of seventeen entries $E(i)$, one first computes the difference $E(1) - E(17)$ between the 1st entry and the 17th entry obtained after a complete rotation of the apparatus. Then, one adds 15/16 of the correction to the 16th entry, 14/16 to the 15th entry and so on, by obtaining a set of 16 entries which are corrected for thermal drift

$$E_{\text{corr}}(i) = \frac{i-1}{16}(E(1) - E(17)) + E(i). \tag{3.20}$$

The fringe shifts are then defined by the differences between each of the corrected entries $E_{\text{corr}}(i)$ and their average value $\langle E_{\text{corr}} \rangle$ as

$$\frac{\Delta\lambda(i)}{\lambda} = E_{\text{corr}}(i) - \langle E_{\text{corr}} \rangle. \tag{3.21}$$

These relevant entries are reported in Table 3.1.

With these entries, Michelson and Morley were then averaging the two sets of data, i.e. over the three noon sessions and over the three evening sessions, and combining the data at angles θ and at $\pi + \theta$ to extract the second harmonic component. The results were then reported graphically in Fig.3.4.

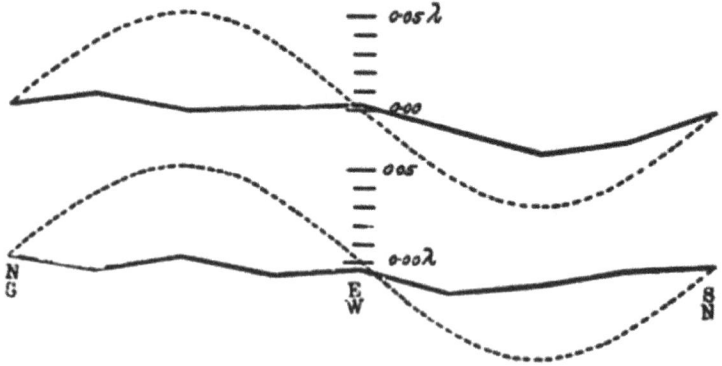

Fig. 3.4   *Michelson-Morley fringe shifts as reported in [1] .*

We observe that, with Miller's definition Eq.(3.21), the fringe shifts are defined up to an overall change of sign. Thus, for instance, the average data (symmetrized over θ and $\pi + \theta$) from the three noon sessions are (-0.006, -0.013, -0.004, -0.006, -0.007, 0.006, 0.017, 0.009) and, as such, differ from those reported in the Michelson-Morley Fig.3.4 by an overall change of sign (but are exactly the same as those shown in Figure 3 of ref. [7]).

Let us now consider the interpretation that Michelson and Morley gave of their data. In their paper, one finds the following statement: "The displacement to be expected was 0.4 fringe" while "...the actual displacement was certainly less than the twentieth part of this". In this way, since the displacement is proportional to the square of the velocity, "...the relative velocity of the earth and the ether is... certainly less than one-fourth of the orbital earth velocity".

The straightforward translation of this upper bound is that the observable velocity, i.e. as obtained from the fringe shifts, is certainly smaller than 7.5 km/s. Though, this estimate is affected by a theoretical uncertainty. The point is that, in Fig.3.4, Michelson and Morley reported their measured fringe shifts together with the plot of a theoretical second-harmonic component (the dashed line). This theoretical wave has an amplitude of 0.05, that they interpret as *one-eight* of the theoretical displacement expected for $v = 30$ km/s, thus implicitly assuming a second-harmonic amplitude $A_2^{\text{class}} = 0.4$.

Instead, as anticipated, the amplitude of the classically expected second-harmonic component is 0.2. Thus 0.4 is the expected shift *after 90 degree rotations* and should be compared with the difference between maxima and minima of the wave. In view of this misunderstanding, their upper bound on the observable velocity should be increased by a factor $\sqrt{2}$.

This conclusion is also consistent with the values obtained from a direct average of the fringe shifts. For the noon observations one finds an experimental amplitude $A_2^{EXP} \sim 0.013$ (and an average phase $\theta_0^{EXP} \sim 135$ degrees, or equivalently $135 + 180$ degrees). Analogously, for the evening sessions one obtains $A_2^{EXP} \sim 0.011$ (and $\theta_0^{EXP} \sim 78$ degrees, or equivalently $78 + 180$ degrees). On the basis of Eq.(3.15), and with a reference classical value $A_2^{class} \sim 0.2$, the mean experimental amplitude ~ 0.012 corresponds to an observable velocity $v_{obs} \sim 7.3$ km/s thus confirming that their 7.5 km/s value cannot be interpreted as a true upper bound.

3.4 A closer look at the Michelson-Morley data

The result and the interpretation of the Michelson-Morley experiment have been (and are still) the subject of endless controversies. For instance, for some time there was the idea [103] that, by taking into account the reflection from a moving mirror and other effects, the predicted shifts would be largely reduced and become unobservable. These points of view are summarized in Hedrick's contribution to the 'Conference on the Michelson-Morley experiment' [102] (Pasadena, February 1927) which was attended by the greatest experts of the time, in particular Lorentz and Michelson. The arguments presented by Hedrick were, however, refuted by Kennedy [3] in a paper of 1935 where, by using the Huygens principle, he re-obtained to order v^2/c^2 the classical result

$$\left[\frac{\Delta\lambda(\theta)}{\lambda}\right]_{class} \sim \frac{D}{\lambda}\frac{v^2}{c^2}\cos 2(\theta - \theta_0) \tag{3.22}$$

where v and θ_0 are respectively the magnitude and the direction of the drift in the interferometer plane. This means that the second-harmonic amplitude was unambiguously predicted as

$$A_2^{class} = \frac{D}{\lambda}\frac{v^2}{c^2}. \tag{3.23}$$

As such, by comparing with Eq.(3.15), in the classical theory the observable velocity v_{obs}, as derived from the fringe shifts, was definitely assumed to coincide with the kinematical velocity v.

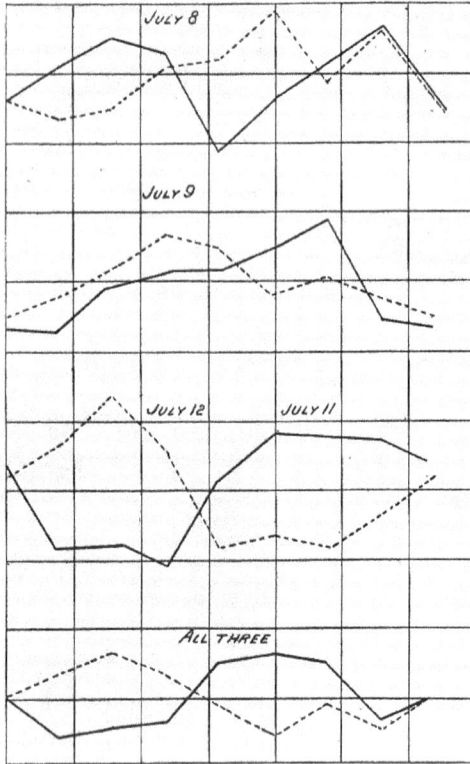

Phil. Mag. S. 6. Vol. 3. Pl. I.

Fig. 3.5   *The Michelson-Morley fringe shifts as reported by Hicks [6]. Solid and dashed lines refer respectively to noon and evening observations.*

Now, in the previous section we have seen that a typical experimental amplitude $A_2^{\mathrm{EXP}} \sim 0.012$ (corresponding to an observable velocity $v_{\mathrm{obs}} \sim 7$ km/s) is obtained from the averaged data. This is much smaller than the classical prediction $A_2^{\mathrm{class}} \sim 0.2$ but is also non negligible when compared to the high precision ± 0.004 of the experimental fringes. Thus it is natural to ask: should this value be interpreted as a typical instrumental artifact (a "null result") or could also indicate a genuine ether-drift effect?

To try to find an answer, let us first recall that, over the years, greatest experts have raised serious objections to the standard null interpretation of the Michelson-Morley data. This point of view was well summarized by Miller in 1933 [7] as follows: "The brief series of observations was sufficient to show clearly that the effect did not have the anticipated magnitude. However, and this fact must be emphasized, *the indicated effect was not zero*".

The same conclusion had already been obtained by Hicks in 1902 [6]: "The data published by Michelson and Morley, instead of giving a null result, show distinct evidence for an effect of the kind to be expected". Namely, there was a second-harmonic effect. But its amplitude was substantially smaller than the classical expectation (see Fig.3.5).

Quantitatively, the situation can be summarized in Fig. 3.6, taken from Miller [7], where the magnitude of the observable velocity measured in various ether-drift experiments is reported and compared with a smooth curve fitted by Miller to his own results as function of the sidereal time.

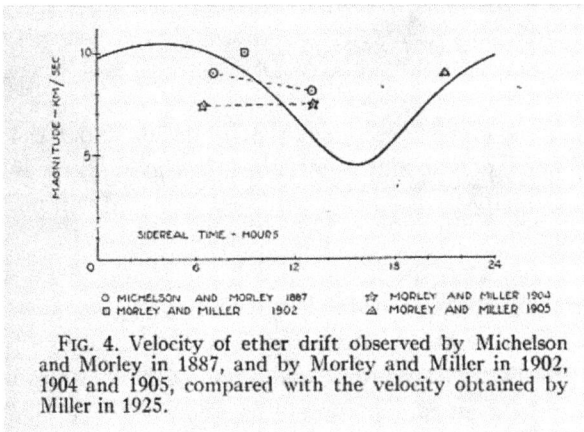

FIG. 4. Velocity of ether drift observed by Michelson and Morley in 1887, and by Morley and Miller in 1902, 1904 and 1905, compared with the velocity obtained by Miller in 1925.

Fig. 3.6 *The magnitude of the observable velocity measured in various experiments as reported by Miller [7].*

For the Michelson-Morley experiment, the average observable velocity reported by Miller is about 8.4 km/s. Comparing with the classical prediction for a velocity of 30 km/s, this means an experimental amplitude

$$A_2^{\text{EXP}} \sim 0.2 \, (\frac{8.4}{30})^2 \sim 0.016 \tag{3.24}$$

which is about twelve times smaller than the expected result. However, this 0.016 value is even larger than the previous value 0.012 obtained from the

July 9 evening

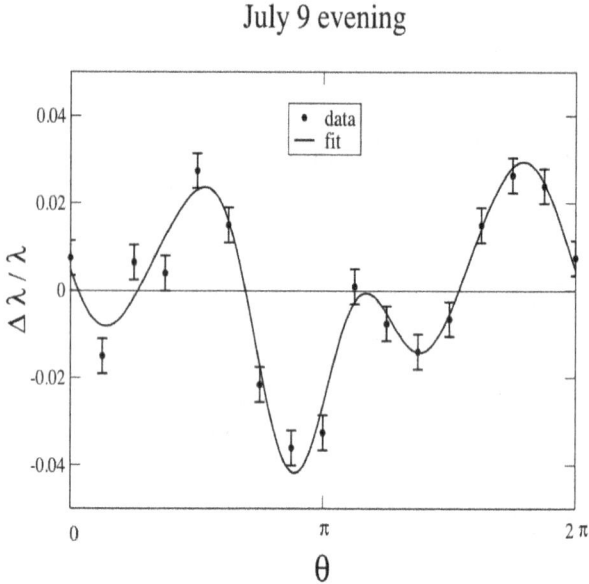

Fig. 3.7 *The fringe shifts for the session of July 9 evening. The fit is performed by including terms up to fourth harmonics. The figure is taken from ref. [105].*

average fringes and confirms the idea that to decide about the significance of the experiment, one should go deeper into the analysis of the data in Table 3.1. This analysis was presented in ref. [38] and will be summarized in the following.

We first observe that the fringe shifts Eq.(3.21) are given as a periodic function, with vanishing mean, in the range $0 \le \theta \le 2\pi$, with $\theta = \frac{i-1}{16} 2\pi$, so that they can be reproduced in a Fourier expansion, see e.g. Fig.3.7.

One can thus extract amplitude and phase of the 2nd-harmonic component by fitting the even combination of fringe shifts

$$B(\theta) = \frac{\Delta\lambda(\theta) + \Delta\lambda(\pi + \theta)}{2\lambda} \qquad (3.25)$$

(see Fig.3.8). This is to cancel the 1st-harmonic contribution originally pointed out by Hicks [6] (which is present in Fig.3.7). Its theoretical interpretation is in terms of the arrangements of the mirrors and, as such, this effect has to show up in the outcome of real experiments. For more details, see the discussion given by Miller, in particular Fig.30 of ref. [7], where it is shown that his observations were well consistent with Hicks' theoretical study. The observed 1st-harmonic effect is sizeable, of comparable magnitude or even larger than the second-harmonic effect. The same conclusion

July 11 noon

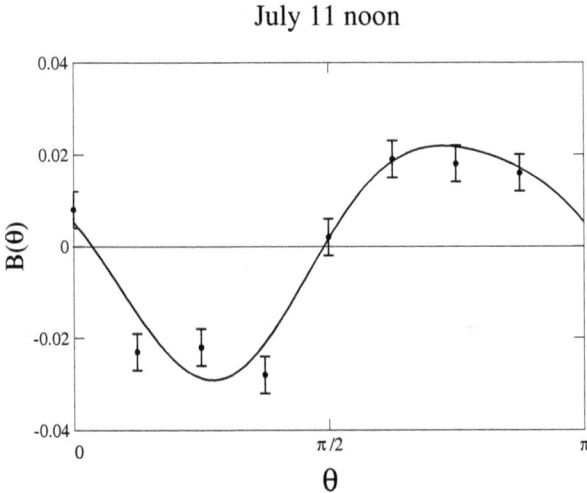

Fig. 3.8 *A fit to the even combination $B(\theta)$ Eq.(3.25). The second harmonic amplitude is $A_2^{\text{EXP}} = 0.025 \pm 0.005$ and the fourth harmonic is $A_4^{\text{EXP}} = 0.004 \pm 0.005$. The figure is taken from ref. [105]. Compare the data with the solid curve of July 11th shown in Fig.3.5.*

Table 3.2 *The amplitude of the fitted 2nd-harmonic component A_2^{EXP} for the six experimental sessions of the Michelson-Morley experiment.*

SESSION	A_2^{EXP}
July 8 (noon)	0.010 ± 0.005
July 9 (noon)	0.015 ± 0.005
July 11 (noon)	0.025 ± 0.005
July 8 (evening)	0.014 ± 0.005
July 9 (evening)	0.011 ± 0.005
July 12 (evening)	0.024 ± 0.005

was also obtained by Shankland et *al.* [47] in their re-analysis of Miller's data. The 2nd-harmonic amplitudes from the six individual sessions are reported in Table 3.2.

Due to their reasonable statistical consistency, one can compute the mean and variance of the six determinations in Table 2 and obtain $A_2^{\text{EXP}} \sim$

0.016 ± 0.006. This value is consistent with an observable velocity

$$v_{\text{obs}} \sim 8.4^{+1.5}_{-1.7} \quad \text{km/s} \tag{3.26}$$

in complete agreement with Miller's Figure 4.

While the individual amplitudes show a reasonable consistency, there are substantial changes in the apparent azimuth θ_0 in the plane of the interferometer. For instance, by taking into account that this is always defined up to $\pm 180^o$, one choice for the experimental azimuths is $357^o \pm 14^o$, $285^o \pm 10^o$ and $317^o \pm 8^o$ respectively for July 8th, 9th and 11th. For this assignment, the individual velocity vectors $v_{\text{obs}}(\cos\theta_0, -\sin\theta_0)$ and their mean are shown in Fig. 3.9.

According to the usual interpretation, the large spread of the azimuths is taken as indication that any non-zero fringe shift is due to pure instrumental effects. However, this type of discrepancy could also indicate an unconventional form of ether-drift where there are substantial deviations from the expected smooth trend associated with slow effects such as the earth rotation and orbital revolution.

For instance, differently from July 11 noon, which represents a very clean indication, there are sizeable 4th- harmonic contributions ($A_4^{\text{EXP}} = 0.019 \pm 0.005$ and $A_4^{\text{EXP}} = 0.008 \pm 0.005$ for the noon sessions of July 8 and July 9 respectively). In any case, the observed strong variations of θ_0 are in qualitative agreement with the analogous values reported by Miller. To this end, compare with Fig.22 of ref. [7] and in particular with the large scatter of the data taken around August 1st, as this represents the epoch of the year which is closer to the period of July when the Michelson-Morley observations were actually performed. Thus one could also conclude that individual experimental sessions indicate a definite non-zero ether drift but the azimuth does not exhibit the smooth trend expected from the conventional picture.

For completeness, we add that the large spread of the θ_0-values might also reflect a particular systematic effect pointed out by Hicks [6]. As described by Miller [7], "before beginning observations the end mirror on the telescope arm is very carefully adjusted to secure vertical fringes of suitable width. There are two adjustments of the angle of this mirror which will give fringes of the same width but which produce opposite displacements of the fringes for the same change in one of the light-paths". Since the relevant shifts are extremely small, "...the adjustments of the mirrors can easily change from one type to the other on consecutive days. It follows that averaging the results of different days in the usual manner is not allowable

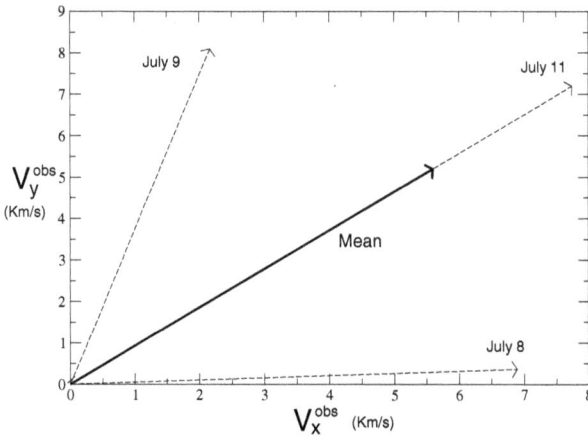

Fig. 3.9 *The observable velocities for the three noon sessions and their mean. The x-axis corresponds to $\theta_0 = 0° \equiv 360°$ and the y-axis to $\theta_0 = 270°$. Statistical uncertainties of the various determinations are ignored. All individual directions could also be reversed by 180°.*

unless the types are all the same. If this is not attended to, the average displacement may be expected to come out zero — at least if a large number are averaged " [6]. Therefore averaging vectorially the fringe shifts from various sessions represents a delicate issue and can introduce uncontrolled errors.

For instance, it can introduce spurious changes in the apparent direction θ_0 of the ether-drift from one session to the other. In fact, a change of sign of the fringe shifts is equivalent to replacing $\theta_0 \to \theta_0 \pm \pi/2$. Clearly, this relative sign does not affect the values of A_2 and this is why averaging the 2nd-harmonic amplitudes is a safer procedure[4].

As a matter of fact, Hicks concluded that the fringes of July 8th were of different type from those of the remaining days. Thus for his averages (in our Fig.1) "the values of the ordinates are one-third of July 9 + July 11 − July 8 and one-third of July 9 + July 12 − July 8" [6] for noon and evening

---

[4]The conventional way of first combining vectorially the fringe shifts of the various sessions, and then extracting the amplitude, was also more recently re-proposed by Handshy [106]. However, his values for these combined fringes do not agree with those reported by Miller and exactly re-obtained by us. For instance, for the noon sessions, the half-period curve shown in Miller's Figure 3 corresponds to (-0.006, -0.013, -0.004, -0.006, -0.007, 0.006, 0.017, 0.009) as previously quoted by us. These values do not agree (apart from a possible overall change of sign) with those reported in Handshy's Figure 4.

sessions respectively. If this were true, one choice for the azimuth of July 8th could now be $\theta_0^{\mathrm{EXP}} = 267^o \pm 14^o$. This would orient the arrow of July 8th in Fig.4 in the direction of the $y-$axis and change the average azimuth from $\langle \theta_0^{\mathrm{EXP}} \rangle \sim 317^o$ to $\langle \theta_0^{\mathrm{EXP}} \rangle \sim 290^o$. Later on [107] Hicks realized that the results of the Michelson-Morley evening observations, although taken in reverse order, were finally printed in direct. This modifies his analysis of the azimuth but leaves totally unaffected the amplitudes extracted from the individual experimental sessions.

In conclusion, our re-analysis supports the claims of Hicks and Miller. Indeed, although the fringe shifts were much smaller than expected, in two experimental sessions (11 July noon and 12 July evening), the second-harmonic amplitude is non-zero at the 5σ level and in other two sessions (July 9 noon and July 8 evening) it is non-zero at the 3σ level. Truly enough, the observed irregular variations of the azimuth may induce (and actually have induced most authors so far) to conclude that such non-zero effects represent typical instrumental artifacts. For this reason, only by disposing of an alternative theoretical model, and after looking at the consistency of different experiments, one will be able to finally conclude whether the small and irregular effects observed by Michelson and Morley have, or not, a genuine physical meaning.

To this end, a new interpretative model and a detailed comparison among different experiments will be presented in Chapt.6. For the moment, however, we prefer to postpone this analysis and instead follow the historical development. Thus we will concentrate, in Chapt.4, on the first formulation of the relativistic effects and on the inception of Einstein's special relativity (from 1889 to 1905) and, in Chap.5, on the traditional view of the classical repetitions of the Michelson-Morley experiment (from 1902 to 1930).

Chapter 4

In 1905 I was convinced that one was not allowed to speak anymore of ether in physics. However, this opinion was too radical. According to the general theory of relativity space is endowed with physical qualities; in this sense, therefore, there exists an ether. But this ether may not be thought as consisting of parts which may be tracked through time. One can thus say that the ether is resurrected in the general theory of relativity, though in a more sublimated form.

A. EINSTEIN, Aether and The Theory of Relativity, Leyden 1920.

4.1 The first Lorentzian version of relativity

The Michelson-Morley experiment had a strong impact on the scientific ambiance. Indeed, apparently, it was challenging the basic tenets of classical physics and/or the very existence of the ether. For the same reason, however, several scientists were induced to sharpen the implications of the electromagnetic theory embodied in Maxwell's equations. From a reconstruction of this original formulation of relativity, see [32,33] and references quoted therein, the important element for Fitzgerald's 1889 [8] original idea of length contraction was his consideration of the *Heaviside ellipsoid*, i.e. the deformation of the electric field configuration associated with a moving electric charge. A similar conclusion was obtained by Lorentz in 1895 and produced the following picture: the length of bodies (held together by electromagnetic forces) moving with uniform velocity v contracts, along the direction of motion, as compared to their equivalent copies at rest in the ether, by a factor

$$\frac{1}{\gamma} = \sqrt{1 - v^2/c^2} < 1. \qquad (4.1)$$

This effect, when taken into account in the computation of the light paths, could explain the null result of the Michelson-Morley experiment.

Nearly at the same time, Larmor in 1900 [12] deduced a complementary phenomenon: the rotation period of a system of two electric charges (negative and positive) whose center of mass moves with uniform velocity v is dilated by a factor $\gamma > 1$ as compared to the rotation period of the same system at rest in the ether. Taking into account both effects, Larmor arrived to what represents the first exact derivation of Lorentz transformations [108, 109]. These were derived by imposing the cancelation of any v^2/c^2 effect of the earth motion in Michelson-Morley experiment.

The importance of this first historical phase was caught by Bell in a chapter on relativity of his book *Speakable and unspeakable in quantum mechanics* [31]. In this essay, he says that, whenever he had the chance to teach special relativity, rather than emphasizing the aspects that were in striking contradiction with the past, he would have tried to concentrate on those elements of continuity[1] with the previous ideas. Bell's presentation has a considerable pedagogical value and will be summarized by us in a simplified form.

To try to reproduce the spirit of the first Lorentzian form of relativity, let us start from the original assumption: the existence of a *preferred reference frame*, say the ether, where Maxwell's equations of electromagnetism are valid. Let us also introduce an observer Σ who is at rest in the ether. His measurements of length and time are performed with suitable rods and clocks whose physical constitution depends on the structure of matter. Let us think matter in a very idealized form as made up of atoms whose fundamental constituents (nucleus and electrons) interact with each other only electromagnetically. Let us also denote by (X, Y, Z, T) the coordinates by means of which the Σ observer parameterizes 3-dimensional space and time.

Let us first consider atomic matter at rest in the ether. In an oversimplified picture (for a more realistic microscopic description see [111]), for

---

[1] In a more recent work De Abreu and Guerra show [110] that, with a weaker formulation of the postulates of relativity, Einstein's view and the original Lorentzian formulation represent merely different perspectives of one and the same physical theory.

the Σ observer electrons move in circular atomic orbits of radius L_0 and period τ_0. Instead, due to the deformations of the electromagnetic field, atomic matter which moves uniformly in the ether exhibits for Σ two peculiar effects. Namely, by defining v' the translation velocity, c the speed of light in the ether frame and $\beta' = v'/c$, Σ observes that:

(1) The linear dimension of the orbits is contracted by a factor $1/\gamma' = \sqrt{(1-(\beta')^2}$ along the direction of motion. This means that the orbits appear to have an elliptic shape where the longitudinal dimension (along the motion) is

$$L_{\text{long}} = L_0\sqrt{1-(\beta')^2} < L_0. \tag{4.2}$$

Instead, the transverse dimensions (perpendicular to the direction of motion) are unchanged, i.e.

$$L_{\text{transv}} = L_0. \tag{4.3}$$

(2) The rotation period is dilated by a factor $\gamma' = \dfrac{1}{\sqrt{(1-(\beta')^2}}$ so that

$$\tau' = \tau_0\gamma' > \tau_0. \tag{4.4}$$

We can then look for a linear transformation, from the (X, Y, Z, T) coordinates of Σ, to (x', y', z', t') coordinates of a new S' observer. The transformation has to be such that matter which is moving (with respect to Σ) is now seen by this new observer precisely as Σ sees the matter which is at rest with respect to him (i.e. circular orbits of radius L_0 with period τ_0). This transformation, which corresponds to the transition to the observer which jointly moves along with matter, is called *Lorentz transformation*. By defining X as the axis of motion, it takes the following form

$$x' = \gamma'(X - v'T); \qquad y' = Y; \qquad z' = Z; \qquad t' = \gamma'(T - \frac{v'X}{c^2}). \tag{4.5}$$

The inverse transformation is obtained by simply replacing the Σ coordinates with the S' coordinates and changing the sign of the velocity, i.e.

$$X = \gamma'(x' + v't'); \qquad Y = y'; \qquad Z = z'; \qquad T = \gamma'(t' + \frac{v'x'}{c^2}). \tag{4.6}$$

For small values of the ratio v'/c these formulas become the Galileo transformations of classical physics: $x' = X - v'T$; $y' = Y$; $z' = Z$; $t' = T$.

Lorentz transformations preserve the form of Maxwell's equations and can explain the null result of a Michelson-Morley experiment in the ideal limit where the velocity of light in Σ coincides with the basic parameter c. In fact, for any moving observer S' light is seen to propagate with the same value $c' = c$ regardless of the magnitude of v' and of the direction of light propagation.

4.2   1905: Einstein's special relativity

The inception of Einstein's special relativity in 1905 [17] has represented one of the most important events in the history of physics. Thousands of books and articles have been written and will be written about Einstein's presentation, his original views, assumptions and so on. Our aim here is not to provide one more account of these aspects but, rather, to keep as much as possible the same Bell's philosophy of an ideal continuity, rather than a radical break with the past. This section is an attempt to explore if, and to which extent, this is possible.

To this end, let us start from a Lorentzian picture and assume that the motion with respect to Σ is definitely unobservable. Then the ether, as a reference frame, could be considered a *superfluous* concept [17].

To be more definite, let us recall that Lorentz transformations belong to the fundamental Lorentz Group so that, in general, the relation between two systems individually connected to a third one by a Lorentz transformation is also a transformation of the group. This means that two systems S' and S'' in uniform translational motion with respect to Σ differ by a suitable *relative* velocity parameter and, in general, by a spatial rotation.

By restricting to the case of a simple one-dimensional motion[2], this means that given a Lorentz transformation from Σ to S'

$$(X, Y, Z, T) \quad \rightarrow \quad (x', y', z', t') \qquad : v' \text{ (with respect to the ether) (4.7)}$$

and given a Lorentz transformation from Σ to S''

$$(X, Y, Z, T) \quad \rightarrow \quad (x'', y'', z'', t'') \qquad : v'' \text{ (with respect to the ether) (4.8)}$$

the relation between S' and S'' can be expressed as

$$(x', y', z', t') \quad \rightarrow \quad (x'', y'', z'', t'') \qquad : v_{\text{rel}} = \frac{v' - v''}{1 - \frac{v'v''}{c^2}}. \qquad (4.9)$$

Therefore if the individual v' and v'' were really unobservable, it should be possible to drop any reference to Σ and derive the relativistic effects only on the basis of the *relative* motion between any pair (S', S'') of observers. This remark could represent Einstein's motivation to abandon a description in terms of the physical effects of the electromagnetic ether and instead derive Lorentz transformations from two *postulates*:

(a) Principle of Relativity (i.e. the complete physical equivalence of all inertial frames in relative, uniform translational motion)

---

[2]We ignore here the subtleties related to the Thomas-Wigner spatial rotation which is introduced when considering two Lorentz transformations along different directions, see e.g. [112–114] .

(b) Independence of the velocity of light (in the *vacuum*) on the state of motion of the emitting source

In this alternative derivation, one just follows the other way around. Those effects (length contraction and time dilation) which were at the base of the original Lorentzian view of relativity, become now a consequence of Lorentz transformations and, so to speak, *dematerialize*. Namely, they loose their original meaning as dynamical modifications of the physical rods and clocks, and become part of the kinematics. Thus, relativity, being now totally disconnected from those original electromagnetic aspects, becomes axiomatic and, in this sense, extendable to any other phenomenon.

This wider perspective can be well summarized by the same Einstein words: "There is no doubt that the theory of relativity, if we regard its development in retrospect, was ripe for discovery in 1905. Lorentz had already observed that the transformations which later were known by his name were essential for the analysis of Maxwell equations and Poincaré had even penetrated deeper into these connections. Concerning myself, I knew only Lorentz' important work of 1895 but not his later work nor the consecutive investigations by Poincaré. In this sense my work of 1905 was independent. The new feature of it was the realization that the bearing of Lorentz transformations transcended its connection with Maxwell equations and was concerned with the nature of space and time in general. The new result was that Lorentz invariance was the general condition for any physical theory" [19].

It should be said that the axiomatization operated by Einstein is not the only aspect of his presentation that should be emphasized. Actually, according to his same words, interpreting relativity as being due to the relative motion eliminates from space and time "the last remainder of physical objectivity" [115]. Namely, besides giving up with the classical idea of a unique absolute space and of a unique absolute time, one also abandons the idea to understand the differences between the various observers in terms of their state of motion with respect to a single reference frame, as in the original Lorentzian formulation.

In spite of this lack of physical objectivity of space and time, however, precise criteria are fixed to compare the judgements of all equivalent observers. This can be well summarized by von Laue's words: "The boldness and the high philosophical significance of Einstein's doctrine consists, in that it clears away the traditional prejudice of one time valid for all systems. Great as the change is, which it forces upon our whole thought, there is found in it not the slightest epistemological difficulty. For in Kant's manner of expression time is, like space, a pure form of our intuition; a schema in which we must arrange events, so that in opposition to subjective and highly contingent perceptions they may gain objective meaning. This arranging can only take place on the basis of empirical knowledge of natural laws. The place and time of the observed change of a heavenly body can only be established on the basis of optical laws. That two differently moving observers, each one regarding himself at rest, should make this arrangement differently on the basis of the same laws of nature, contains no logical impossibility. Both arrangements have, nevertheless, objective meaning since there may be deduced exactly from each of them by the derivative transformation formulae that arrangement valid for the other moving observer" [115].

Nevertheless, by assuming that the ether-drift experiments give always true null results, it remains a substantial phenomenological equivalence of the two formulations. This was well summarized by Ehrenfest in his lecture 'On the crisis of the light ether hypothesis' (Leyden, December 1912): "So, we see that the ether-less theory of Einstein demands exactly the same here as the ether theory of Lorentz. It is, in fact, because of this circumstance, that according to Einstein's theory an observer must observe exactly the same contractions, changes of rate, etc. in the measuring rods, clocks, etc. moving with respect to him as in the Lorentzian theory. And let it be said here right away and in all generality. As a matter of principle, there is no experimentum crucis between the two theories".

This phenomenological equivalence reflects the basic group properties of Lorentz transformations and does not depend on the introduction of new forms of interactions (such as weak and strong interactions) which were unknown in 1905. Therefore, even though most authors are perfectly satisfied by explaining relativistic effects as simple consequences of relative motion, it is also true that, if one starts to think about their ultimate physical interpretation, one should agree with Pauli that "the contraction of a rod is a very complicated process and by no means elementary" [116]. In this sense, a satisfactory synthesis between *kinematical* and *dynamical* interpretations of relativity may consist in asserting that "relativistic effects like length contraction and time dilation are in the last analysis the result of structural properties of the quantum theory of matter" [33] once the fundamental equations of the theory are symmetric under the Lorentz Group. As such, one can always frame relativity within a suitable form of ether which is not purely electromagnetic.

Rather, we emphasize that the second Einstein postulate, i.e. the independence of the velocity of light in vacuum from the state of motion of the emitting source, is characteristic of wave propagation in a medium and, in this sense, as remarked by Pauli, "it represents the true essence of the old ether point of view" [116].

4.3 Einstein and the ether

In spite of their substantial phenomenological equivalence, there is no doubt that in the "contest" between Lorentzian relativity and Einstein's axiomatic formulation, the latter has prevailed. Certainly this might be attributed to the wider perspective of Einstein's view and also to his other contributions (the photoelectric effect, Brownian motion, specific heats, general relativity etc.) having raised him to a quite unprecedented level among scientists.

There is, however, another reason for this preference, which has its roots in something that made an even greater sensation in his 1905 article : the phrase "The introduction of a luminiferous ether will prove to be superfluous" [17]. This was the fundamental novelty: Einstein's declaration that he could do without the ether, and which distinguished him from any other scholar.

After this, a form of resentment started to grow amongst some leading members of German science who instead were convinced that the ether was essential in order to fully understand electromagnetic and gravitational phenomena. This hostility, especially in that difficult period for Germany after the First World War which culminated in the advent of Nazism, transcended the purely scientific and started to acquire extremely ugly tones, something undoubtedly further exacerbated by Einstein's pacifism and Jewish origin. Because of these deplorable attacks on Einstein himself, a negative image started to be attributed to anything connected with the ether, including the Lorentzian formulation of relativity.

We believe that this has represented an overall loss for physics, not least because during the next few years Einstein's point of view was to change. This evolution is well-documented by Ludvik Kostro in his book "Einstein and the ether" [117] where the author describes Einstein's ideas and vicissitudes through many original documents which cover a period of several decades. Besides the reconstruction of Einstein's thought, the book offers a beautiful slice of history (and sociology) of science and very interesting perspectives on early 20th century German society. We give a brief summary of it here.

From Kostro's research, it first of all emerges that during Einstein's early years in Bern, besides reading Lorentz' 1895 article and Poincaré's 1902 book "La Science et l'Hypothese", he was also influenced by the thoughts of Mach and of the chemist and naturalist Ostwald. While Mach accepted the concept of ether as a necessary medium to explain the local inertia of bodies by the presence of very far and large masses, Ostwald rejected the very idea, arguing that its physical properties could not be measured directly: "What we measure is energy ... therefore there is no need to look for a carrier if we detect energy somewhere".

Nevertheless, in his 1905 article, Einstein adopted a somewhat weaker term ("superfluous") thus intending to imply that he was not in principle denying the possible existence of the ether. However, this is just what he then proceeded to do. Indeed, shortly after this, his original cautiousness was replaced by much stronger statements. For instance, in 1910, he was writing that to reconcile electromagnetism with the principle of relativity, "the first thing to do was to give up with the ether". But as we have seen in the essential equivalence of his approach with the Lorentz formulation, this explicit denial of the ether had no real justification. It brought him into serious dispute with several colleagues, in particular with Philipp Lenard, an important experimentalist who had been awarded the 1905 Nobel Prize for his measurements on the photoelectric effect, the same effect, in this case for its theoretical interpretation, for which Einstein would also win the Nobel Prize in 1921.

At the beginning, precisely because of their common interest in the photoelectric effect, Einstein and Lenard held each other in mutual esteem. This situation, however, was shortly to change with Lenard's 1910 work "Ether and matter" where, without attacking Einstein directly, he nevertheless defended the idea that ether was important in explaining many electromagnetic and gravitational phenomena. According to Lenard, the ether should have been thought not as a continuum but as made up of very small rotating parts and this could explain why material bodies could pass through without any observable friction.

This idea, which as we have seen from Maxwell's conference in Chap.2 was far from new, provoked a strong reaction from Einstein. He wrote to Laub, then Lenard's assistant in Heidelberg, in language highly critical of Lenard. Kostro doesn't say if Lenard knew about this letter. But if so it was certainly not immediately. In fact, three years later in 1913, he was still writing to Sommerfeld about the possibility of inviting Einstein to hold a new chair of theoretical physics in Heidelberg. In view of future developments, however, he might well have become aware of it later.

In 1914, just before the outbreak of the first World war, Einstein was offered a position at the Prussian Academy of Science in Berlin. Newly arrived, he was asked to write a short article for the local press. Here, he was still writing that disposing with the ether had to be considered as one of the most important achievements of the theory of relativity.

However, with the transition from special relativity to general relativity in 1916, Einstein's ideas were to change. In his view of gravitational phenomena, space-time now became a deformable entity whose metric properties are determined by the matter and energy contained therein. Clearly, this new vision has many analogies with the physics of elastic media and leads naturally to the concept of some form of ether. For instance, about fifty years before, Riemann [118, 119] had already had the intuition to connect physics and geometry through an ether whose resistance to deformations could be read in the metric properties of space.

This change of perspective was noted by Lenard in his 1917 work "Principle of Relativity, Ether, Gravitation". Einstein's general relativity was just renaming ether as "space". The paradoxical aspect was that this new theory could not exist without an ether, even though the self-same author of the theory was explicitly denying its existence.

Einstein replied in the form of a dialogue between a critic (who should have been Lenard) and a relativist (Einstein himself) but from his arguments, it seemed that he shared the same basic idea. The phenomena in a gravitational field implied that space had physical properties and, therefore, were equivalent to some form of ether.

Because of Lenard's criticism, his close contacts with Hermann Weyl and his correspondence with Lorentz, who remained true to his convictions until his death in 1928, Einstein thus arrived at a new vision. This rethinking may have been further facilitated by a certain degree of insouciance: in fact, his fame had become immense after the observation of the deflection of light in the 1919 eclipse expedition lead by Eddington.

This change of heart is well-documented in the "Morgan Manuscript" (1920) and in his Leyden Lecture (1920): "In 1905 I was convinced that one was not allowed to speak anymore of ether in physics. However, this opinion was too radical. According to the general theory of relativity space is endowed with physical qualities; in this sense, therefore, there exists an ether ... According to the general theory of relativity space without ether is unthinkable; for in such space there not only would be no propagation of light, but also no possibility of existence for standards of space and time (measuring-rods and clocks)... But this ether may not be thought as consisting of parts which may be tracked through time ... One can thus say that the ether is resurrected in the general theory of relativity, though in a more sublimated form".

Such attempts to clarify his views, rather than favoring a rapprochement, were interpreted as a symptom of weakness, a form of scientific revisionism. For instance Gehrcke, in his essay "Criticism and history of the new theories of gravitation", tried to demonstrate contradictory aspects in the general theory of relativity and accused Einstein of plagiarism concerning the problem of the Mercury perihelion. Wieland, who was not a physicist but one of the founders of the German Association for the Preservation of Science, disposed of a great deal of money offering rewards to anyone who could write attacking Einstein's theory or speak against him at gatherings. On August 6 1920 he wrote an article for a popular journal entitled "Einstein's theory of relativity as scientific mass suggestion", in which he reiterated Lenard's criticisms and Gehrcke's accusations of plagiarism.

Einstein was defended by Max von Laue (1914 Nobel laureate for his research on X rays) with an article published in the same journal on August 11. As a reply, Wieland and Gehrcke invited von Laue to a public debate to be held on August 24 at the Berlin Philharmonic. Several members of Berlin Academic Society were present at this meeting, including Walter Nernst. Von Laue did not accept the challenge but Einstein went instead and applauded with amusement every time he was attacked. These criticisms had no real scientific basis and were mostly centered on the apparent contradictions in the theory of relativity which seemed to reflect "the chaos of thought typical of dadaists". The session finally ended with the usual accusations of plagiarism.

Probably because he did not feel himself as being seriously attacked, Einstein did not reply at the public meeting. However, he did reply three days later with an article in the same journal, "My response to the limited liability company of antirelativists". In this article, after emphasizing that all the most important scientists (Lorentz, Planck, Sommerfeld, von Laue, Born, Larmor, Eddington, Debye, Langevin and Levi-Civita) had acknowledged the value of his theory, he attacked Lenard directly. Whereas he declared his appreciation of Lenard's work as an experimentalist, "however in theoretical physics, so far, he has done nothing and his objections to general relativity are so superficial that I do not care to reply".

The dispute with Lenard continued in the following month at the National Congress of German Naturalists in Bad Nahueim. Most of the country's leading scientists were in the audience and many talks were dedicated to relativity. In particular, Hermann Weyl emphasized that the metric of the space-time continuum could be taken as a definition of the state of the ether in the presence of gravitational fields. At the end of Weyl's talk, Lenard took the floor. After expressing his pleasure that Weyl's theory of gravitation was again referring to the ether, he did not miss the opportunity to repeat his harsh criticisms of Einstein, who answered in kind.

After Bad Nahueim, Lenard did everything to weaken Einstein's prestige and ridicule his theory, as can be seen from the preface to his new book on the ether. Entitled "Admonishment to German researchers", it warned German scientists not to be influenced by what was claimed to be nothing but a pile of hypotheses which untimely had been defined as a "theory". Another important German scientist, Johannes Stark (the 1925 Nobel laureate for physics) joined Lenard. In 1922, in an article entitled "The present crisis of German physics", he wrote: "If only Einstein had left with the mathematicians and philosophers at the very beginning. Perhaps, then German physics could have been preserved from that poison which has paralyzed its thought".

With time, Einstein's life became increasingly difficult, again due to his pacifism and Jewish origins. Take for example the 1931 book "A Hundred authors against Einstein" where articles and speeches of his opponents were collected. The situation became critical when Hitler took power on January 30 of 1933. On March 20 of the same year, while Einstein was visiting the United States, his summer cottage was searched for weapons which, according to somebody, had been left by the communists. Eight days later, Einstein came back to Europe but never more returned to Germany. For some time he settled in Belgium, watched by Belgian secret service, and on October 17 of the same year he went to live indefinitely in Princeton.

So it should not be surprising that for all these hardships, related to his new vision of the ether and his bitter disputes with Lenard, Stark and the others, Einstein felt in the end a form of rejection. Maybe it was for this reason that his idea of "sublimated ether" was not further seriously considered and developed after he moved to United States. The only relevant reference is contained in his 1938 book with Infeld [120]. After recalling the history of the problem, he concluded that "Our attempts to discover the properties of the ether led to difficulties and contradictions. After such bad experiences, this is the moment to forget the ether completely and to try never to mention its name".

Chapter 5

*In the literature of physics there are to be found reports, by serious
investigators, of the occurrence of effects which could not be
reproduced, since further tests led to negative results. A well known
example from recent times is the unexplained positive result of
Michelson's experiment observed by Miller (1921-1926) at Mount Wilson,
after he himself (as well as Morley) had previously reprouced
Michelson's negative result. But since later tests again gave negative
results it is now customary to regard these latter as decisive, and to
explain Miller's divergent result as due to "unknown sources of error".*

K. POPPER, The Logic of Scientific Discovery, 1934.

5.1 After Michelson-Morley: Morley—Miller (1902-1905)

The first repetition of the Michelson-Morley experiment was performed
by Morley and Miller in the years 1902-1905. An important stimulus for
this new series of measurements came from the conversations that the two
physicists had with Lord Kelvin at the International Congress of Physics
held in Paris in the occasion of the Exposition of 1900. In his talk, Kelvin
had exposed some models of the ether and discussed the importance of the
Michelson-Morley experiment whose unexplained outcome was deserving
more accurate checks. For this reason, he strongly urged Morley and Miller
to repeat the experiment with an improved apparatus.

After Paris, Morley and Miller started with the construction of a new interferometer designed especially to test the hypothesis of the Fitzgerald-Lorentz contraction [7]. The base of the instrument was made of pine wood and the total optical path was about three times longer than the one adopted in the original 1887 experiment. The instrument was mounted in the basement of the Case School of Cleveland and the first series of observations in 1902 and 1903 consisted of 505 turns of the interferometer. A small positive effect was observed as indicated by the square in Miller's Figure 4 reported by us in Chap.3. The small observed velocity, which was comparable to that observed in the 1887 experiment, supported the conclusion that if the reduction was due to the hypothetical contraction, pine was affected by the same amount as sandstone.

The unwanted modifications of the wooden structure, due to humidity and temperature changes, induced, however, Morley and Miller in 1904 to build an entirely new apparatus of steel. The purpose was to secure structural symmetry and utmost rigidity [7]. The optical flat surfaces were made by O. L. Petitdidier of Chicago and proved to be exceptionally perfect. With the arranged system of reflecting mirrors, the total light path, going and returning, reached a length 2D=6406 centimeters, equal to 112,000,000 wavelengths of the acetylene light used in the experiment. By restricting to the earth orbital motion, and according to the classical theory, this figure would then correspond to fringe shifts with a typical second harmonic amplitude

$$A_2^{\text{class}} = \frac{D}{\lambda} \frac{(30 \text{ km/s})^2}{c^2} \sim 0.56 \qquad (5.1)$$

The first observations with this improved apparatus were made in July 1904 and consisted in 260 turns of the interferometer [7]. The procedure adopted for the analysis of the data was based on assuming that the earth motion, relevant for the hypothetical ether-drift effect, had to be obtained by combining the motion of the solar system relative to nearby stars, i.e. toward the constellation of Hercules with a velocity of about 19 km/s, with the annual orbital motion ("We now computed the direction and the velocity of the motion of the center of the apparatus by compounding the annual motion in the orbit of the earth with the motion of the solar system toward a certain point in the heavens...There are two hours in each day when the motion is in the desired plane of the interferometer" [121]).

The observations at the two times (about 11:30 a.m. and 9:00 p.m.) were, therefore, combined in such a way that the presumed azimuth for the morning observations coincided with that for the evening ("The direction of the motion with reference to a fixed line on the floor of the room being computed for the two hours, we were able to superimpose those observations which coincided with the line of drift for the two hours of observation" [121]). However, the observations for the two times of the day gave results having nearly opposite phases. When these were combined, the result was nearly zero. For this reason, the value then reported of an observable velocity of 3.5 km/s is incorrect and does *not* correspond to the actual results of the individual measurements. The error was later understood and corrected by Miller who found that the two sets of data were each indicating an effective velocity of about 7.5 km/s (see Figure 11 of Miller's paper [7]).

Finally, in 1905, the interferometer was mounted in a site on Cleveland Heights which was at an altitude of about 285 meters. The observations, made in July, October and November, 1905, and consisting of 230 turns of the interferometer, showed a small but definite effect with an observable velocity of about 8.7 km/s.

Thus, summarizing, the average *observable* velocities, for the entire period 1902-1905 of the Morley-Miller experiment, are those shown in Miller's Figure 4 (see our Chapt. 3) and lie approximately between 7 and 10 km/s or

$$v_{\text{obs}} \sim (8.5 \pm 1.5) \text{ km/s}. \tag{5.2}$$

5.2 Miller 1920−1925

The advent of Einstein's special relativity in 1905, of his general theory of relativity in 1916 and the crucial observation of the deflection of light in 1919, in the eclipse expedition lead by Eddington, determined radical changes in the scientific views. On the one hand, the idea that it was impossible to detect any ether-drift effect was becoming more and more natural. On the other hand, experimental evidence for both the undulatory and corpuscular aspects of radiation was substantially modifying the consideration of an ether and its logical need for the physical theory.

As emphasized by Swenson [98], Miller's figure of scientist should be framed in this rapidly changing scientific ambiance. In particular, for a keen experimentalist as he was, "the half-truths and false-hoods about the Michelson-Morley experiment which had become common currency" [98] were particularly disturbing and became a strong motivation for a new series of ether-drift observations.

Miller's experiments and his data analysis cover a period of about 13 years, from 1920 to 1933. The many facets of this research and of its reception in United States and Europe represent a fascinating chapter of history of science that has been described in great detail by Swenson [98] and Lalli [122].

The surprise produced by Miller's various announcements for a small but non-zero ether-drift of about 8÷10 km/s was that he was using the same apparatus used by Morley and Miller in their measurements of 1905. As explained above, before Miller's reanalysis, those measurements were believed to yield the much lower value of 3.5 km/s. Hence, for his long-time experience with this type of experiments and his important role in the American Science (in 1921 he became also President of the American Physical Society), his claims could not be easily dismissed. Thus other scientists (Kennedy in 1926, Illingworth in 1927, Joos in 1930 ...) decided to undertake independent checks. In these other trials, apparently, Miller's results were not confirmed but the real motivations for the discrepancy could not be easily understood.

For this reason, the resulting controversy and its sociological aspects represented for Popper [123] (at the time of 1934) a paradigmatic case: "In the literature of physics there are to be found reports, by serious investigators, of the occurrence of effects which could not be reproduced, since further tests led to negative results. A well known example from recent times is the unexplained positive result of Michelson's experiment observed by Miller (1921-1926) at Mount Wilson, after he himself (as well as Morley) had previously reproduced Michelson's negative result. But since later tests again gave negative results it is now customary to regard these latter as decisive, and to explain Miller's divergent result as *due to unknown sources of error*".

The debate finally ended, after Miller's death, with a critical paper by Shankland and collaborators in 1955 [47] which is usually believed to have definitely closed the question by identifying the cause of the effect as mainly due to temperature variations within Miller's interferometer. However, as we shall see, after the discovery of the Cosmic Microwave Background and by disposing of an alternative model for the interpretation of the data, the conclusion that one is really dealing with *spurious* effects is far from obvious (see Chapt.6). To better understand the various aspects, however, let us follow the historical development by first starting with the early 1920−1925 period.

As anticipated, the main motivation for Miller's new research was that "the theory of relativity postulates an exact null effect from the ether-drift experiment which had never been obtained in fact" [7]. A small but definitely non-null ether-drift was thus a challenge to Einstein's relativity. As reported by Lalli [122] (and somehow differently from Swenson's account), in the early 1920s the US scientific community did not share Miller's opinion. Thus, when Miller asked George E. Hale, the director of Mount Wilson Observatory, to repeat the experiment there (in order to check the dependence on altitude), Hale was hesitant and decided to consult Michelson. According to Michelson, any possible non-null effect was almost certainly due to temperature changes thus, in his reply to Miller, Hale agreed to host the experiment but refused to provide financial support.

It was with the support of the Carnegie Institution of Washington that the same Morley-Miller interferometer was set up on Mount Wilson in March, 1921, on the grounds of the Mount Wilson Observatory, on Rock Crusher Knoll or "Ether Rock" as it came to be called, near the site of the 100-inch telescope, at an altitude of about 1750 meters [7]. The sides of the house were enclosed with sheets of corrugated iron, except at a height of the apparatus where on all sides there were windows of white canvas cloth which could be opened on all sides. In this way, one could also establish a free circulation of air to secure equalization of temperature with the outside air [7]. In order to secure sufficient darkness for the observation of the fringes in the daytime, sheets of thin black paper were placed over the canvas windows. Common and precision thermometers were hung on each side of the house and were read at the beginning and end of each set of observations. A barograph and a thermograph were carried at all times on the interferometer itself. These general arrangements were also applied to all subsequent measurements [7].

Observations started on April 8 and continued till April 21, 1921 and consisted of 67 sets for a total number of 350 turns of the interferometer. This first series of measurements (with the steel base of the interferometer) indicated a small periodic effect with an average second harmonic amplitude of about 0.04 [47]. When compared to the classical expectation of 0.56 for $v=30$ km/s, this was indicating an observable velocity of about 8 km/s. These measurements were also accompanied by a disturbance which was periodic with a full turn of the interferometer [122], i.e. a 1st-harmonic effect.

It is interesting that precisely at the same time, May 1921, Einstein was visiting United States to deliver four lectures on relativity at the Institute for Advanced Study in Princeton. Although Miller had not yet announced his results, still, at the end of one of these lectures, someone in the audience (perhaps Ludvik Silberstein [122]) asked Einstein if he was aware of Miller's claim. To this question, Einstein replied by saying "Subtle is the Lord but malicious He is not"[1] implying that he was not giving any chance to the possibility of a non-null ether-drift. Nevertheless, in spite of his scepticism, Einstein went to Cleveland on 25 May 1921 and, apparently, talking to Miller stressed the importance of repeating the experiment in different periods of the year [122].

New measurements in December 1921 (with the concrete base of the interferometer) were reported by Miller in April 1922 [124] and showed "a definite displacement, periodic in each half revolution of the interferometer, of the kind to be expected, but having only one tenth of the presumed amount", i.e. an effective velocity of about 9 km/s.

Since the 1921 experiments gave essentially the same results, Miller concluded that magnetostriction was marginal and could not be the cause of the small observed effects. After the December 1921 test, he transferred the apparatus to Cleveland to perform a long series of other checks, mostly on the relevance of temperature changes. His final conclusion was that, under the actual experimental conditions, the observed changes could not possibly be produced by temperature variations [122]. His general argument, also reaffirmed in ref. [7] and in ref. [125], was that the ether-drift derives from regularly periodic variations in the position of the fringe system. Thus, to reproduce the observed pattern, *temperature should have been changing according to the orientation of the interferometer* and could not be due to accidental variations in the temperature of the room. In fact, this other type of effect would have been eliminated by averaging over a long series of measurements.

After two years, Miller returned to Mount Wilson at a new site which was better shielded by wind currents. On 4, 5 and 6 of September 1924 he carried out ten sets of observations for a total of 136 turns of the interferometer which exhibited the same, small positive effect with a magnitude corresponding to about 9 km/s. New observations in March and April 1925 were apparently confirming the result. However, as we will explain later, the interpretation of the measurements in terms of a definite earth cosmic motion was by no means clear. About these various aspects, he reported at a meeting of the National Academy of Science in April 1925.

In his relation, although lacking a complete, theoretical interpretation of the results, Miller emphasized the *observational strength* of his measurements because the effect was checked in many ways and persisted since

---

[1]This famous sentence later became the title of Abraham Pais' scientific biography of Albert Einstein.

1921 [122]. This bold announcement induced several physicists (notably Eddington, Lorentz, Thirring, Einstein...) to get critically involved into the problem. In spite of the criticisms, Miller went on with his measurements and, in July of the same year 1925, published a brief report stressing that the full-period effect (a 1st harmonic) was a necessary geometric consequence of the adjustment of the mirrors when dealing with fringes of finite width [122]. As anticipated in our analysis of the Michelson–Morley experiment, this point had already been stressed by Hicks in 1902 [6].

5.3   Tomaschek 1924

By assuming an exact null result of the ether-drift experiments with terrestrial sources of light, in his 1920 paper "On ether and primordial ether", Philipp Lenard argued that the ether carrying quanta of light emitted by terrestrial sources should be considered as taking part of the earth motion. The situation, however, might be different when using extraterrestrial light, e.g. from the sun, the moon and the fixed stars. Therefore, in an experiment with such sources of light some positive outcome might be expected. Such experiment was realized by Rudolph Tomaschek in 1923 in Heidelberg [126, 127].

An exact null result of the experiments with terrestrial light was presupposed and, thus, the apparatus was designed to show the possibly differential behavior of light coming from extraterrestrial sources (experimental ray) or from a terrestrial source (reference ray). In the most refined version of the experiment, the optical path had a length of 860 cm. and the optical devices were arranged in such a way to adjust the trajectory of both rays to each other. Dealing with extraterrestrial light the apparatus was held fixed in the laboratory with the two arms placed in the East-West and North-South directions. The rotation of the interferometer was then replaced by the rotation of the earth, given that observations were made at different times of the day.

After having adjusted the central interference fringe of the reference light and after having adapted the brightness of the two rays, measurements could be started. Often, the presence of air flows was producing an instability of the fringe system. Therefore experimental data were collected when the environment was as quiet as possible. A typical linear drift of about 0.05 fringe per minute was also observed so that a single measurement (taking approximately 8 seconds) had an intrinsic uncertainty of about ±0.006.

The classical prediction for the 2nd harmonic amplitude was

$$A_2^{\text{class}} = \frac{D}{\lambda}\frac{(30 \text{ km/s})^2}{c^2} \sim 0.15 \qquad (5.3)$$

and the expected time dependence of the fringes due to (the daily variation of the projection of) the earth orbital velocity was

$$\Delta(t) = A_2^{\text{class}}(1 - \sin^2 t(1 + \sin^2 \phi)) \qquad (5.4)$$

where t is the angle of the hour (for 12h00 one has $t = 0$) and $\phi \sim 49$ degrees is the latitude of the laboratory in Heidelberg.

Following Morley and Miller, Tomaschek was also considering the combination of the earth orbital motion with the motion of the solar system as a whole relatively to the surrounding fixed stars. The two effects were then expected to produce overall shifts between +0.08 and +0.19.

By using light from the sun, from the moon, from Jupiter, Sirius, Vegas and Arthur, at various hours of the day and in different periods of the year, the experimental results were summarized by Tomaschek as follows: "In summary, a shift of 0.1÷0.2 parts of a fringe would be expected. The experiments show that a shift of this size has not occurred. The deviations observed amounted to 1/8 of this expected size at most, this value ranging already within the limits of the instrumental errors" [127].

Therefore, Tomaschek's final conclusion was:"Based on our original viewpoint and given our arrangement, the experiments support the conclusion that light quanta from extraterrestrial sources are no longer in connection with the primordial ether of the universe but, instead, that their speed equals the speed of light in relation to the ether of the earth, which we have assumed at rest with respect to the surface of the earth" [127].

Before ending this section, we should translate Tomaschek's words into an upper limit for the observable velocity. This is not free of ambiguities since one could take as a maximum observed shift the value 0.2/8=0.025, or 0.15/8∼ 0.019 or 0.1/8=0.0125. Here, we shall adopt the most conservative estimate by Shankland et al. [47] $A_2^{\text{EXP}} \sim 0.010$. Then, by taking the mentioned ±0.006 accuracy, we obtain an observable velocity

$$v_{\text{obs}} \sim 30 \text{ km/s} \sqrt{\frac{0.010 \pm 0.006}{0.15}} \sim 7.7^{+2.1}_{-2.8} \text{ km/s}. \qquad (5.5)$$

5.4   Miller 1925−1926

The reactions to his first announcement of April 1925 induced Miller to perform a new series of measurements in August and September 1925 with the idea of presenting old and new results at the joint meeting of the American Physical Society and of the American Association for the Advancement of Science in December of the same year in Kansas City. His goal was to reconstruct the earth cosmic motion from the fringe shifts observed in the laboratory. To this end, the observations were equally distributed over the twenty-four hours of the day to determine the curve of diurnal variations. The procedure was then repeated for a few days and the values collected at the same sidereal time were grouped to form the average corresponding to the given epoch. The conclusion of this study are reported in his final article of 1933 [7]. Miller's paper represents the most comprehensive review of these experiments and the main aspects will be summarized in the following.

In spite of the many evidences that Miller produced, the conclusions of his work have never been accepted by the physical community for two main reasons. On the one hand, the typical velocity of 8÷10 km/s, deduced from the magnitude of the fringe shifts, had no obvious interpretation. On the other hand, his observations did not fit with the interpretative model used to compare data obtained at different epochs of the year.

To separate the two aspects, let us first concentrate on the magnitude of the effect and consider the only set of Miller's data which is known from the literature. This sets corresponds to 20 turns of his apparatus and is explicitly reported in Fig.8 of ref. [7]. To analyze these data, we have followed the definite procedure to subtract the linear drift which has been explained in Chap.3. The resulting symmetric combinations of fringe shifts

$$B(\theta) = \frac{\Delta\lambda(\theta) + \Delta\lambda(\pi + \theta)}{2\lambda} \tag{5.6}$$

are reported in our Table 5.1.

We have then fitted these data by including both 2nd and 4th harmonic terms. Notice that we do not perform any averaging of data obtained from different turns of the interferometer. For our global fit, to estimate the accuracy of the various determinations, we have followed ref. [47] and adopted a nominal uncertainty ±0.050 for each entry of Table 5.1. From the fit, where the 4th harmonic is completely consistent with the background ($A_4^{\mathrm{EXP}} = 0.004 \pm 0.012$), we have obtained a chi-square of 130 for 157 degrees of freedom and the following values

$$A_2^{\mathrm{EXP}} = 0.061 \pm 0.012 \qquad\qquad \theta_0^{\mathrm{EXP}} = 24^o \pm 7^o. \tag{5.7}$$

Table 5.1   *The symmetric combination of fringe shifts $B(\theta)$ Eq.(5.6) at the various angles θ for the set of 20 turns of the interferometer reported by Miller in Figure 8 of ref. [7]. For our global fit, following ref. [47], the nominal accuracy of each entry has been fixed to ±0.050. The values are taken from ref. [38].*

Turn	0°	22.5^o	45°	67.5^o	90°	112.5^o	135°	157.5^o
1	+0.091	+0.159	+0.028	+0.047	−0.034	−0.116	−0.147	−0.028
2	−0.025	+0.063	+0.050	+0.088	−0.075	−0.038	+0.000	−0.063
3	+0.022	+0.103	+0.084	+0.016	−0.053	−0.072	−0.091	−0.009
4	+0.034	−0.009	−0.053	−0.047	−0.041	+0.016	+0.022	+0.078
5	+0.169	+0.081	+0.044	−0.044	−0.081	−0.169	−0.056	+0.056
6	−0.025	+0.025	+0.025	+0.025	+0.025	−0.025	−0.025	−0.025
7	+0.081	+0.094	+0.056	+0.069	−0.119	−0.106	−0.094	+0.019
8	+0.066	+0.072	−0.022	−0.066	−0.059	−0.003	+0.003	+0.009
9	+0.041	+0.084	+0.078	+0.022	−0.134	−0.141	+0.003	+0.047
10	+0.016	+0.072	+0.078	−0.016	−0.009	−0.003	−0.047	−0.091
11	+0.009	+0.053	+0.097	−0.009	−0.116	−0.072	+0.022	+0.016
12	+0.022	+0.016	+0.059	+0.003	−0.053	−0.009	−0.016	−0.022
13	+0.000	+0.063	+0.025	+0.038	+0.050	−0.038	−0.075	−0.063
14	−0.034	+0.047	+0.078	+0.009	−0.009	−0.028	−0.047	−0.016
15	+0.113	+0.125	+0.138	+0.000	−0.088	−0.125	−0.113	−0.050
16	+0.025	+0.050	+0.025	+0.050	−0.025	−0.050	−0.025	−0.050
17	+0.000	−0.012	−0.025	+0.063	+0.000	−0.012	−0.025	+0.013
18	+0.044	+0.050	+0.019	−0.019	−0.056	−0.044	−0.031	+0.031
19	+0.053	+0.059	+0.016	−0.028	−0.022	−0.066	−0.009	−0.003
20	+0.059	+0.041	+0.122	+0.003	−0.066	−0.084	−0.053	−0.022

Here errors correspond to the overall boundary $\Delta\chi^2 = +3.67$, as appropriate[2] for a 70% C. L. in a 3-parameter fit [129]. The above average amplitude $A_2^{\mathrm{EXP}} = 0.061 \pm 0.012$ corresponds to an observable velocity $v_{\mathrm{obs}} \sim 9.9 \pm 1.1$ km/s.

We emphasize that the fitted A_2 Eq.(5.7) is only 20% larger than the nominal accuracy ±0.050 of each entry but the data are distributed in such a way to produce a 5σ evidence for a non-zero 2nd harmonic. This shows the inconsistency of the criticism [128] that this sample of data has no statistical significance. Concerning other objections raised in ref. [128] we address the reader to ref. [38].

To give an idea of the characteristic scatter of the results obtained from the various turns, we have reported in Figs. 5.1 and 5.2 the experimental values of the azimuth θ_0 and of the 2nd harmonic A_2 for the 20 rotations.

---

[2]This probability content assumes a Gaussian distribution as for typical statistical errors.

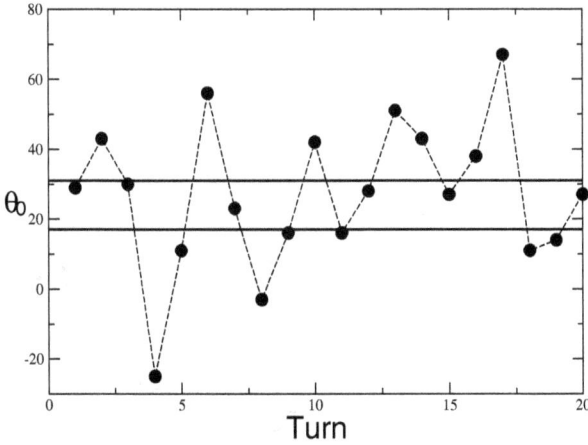

Fig. 5.1 *The azimuth (in degrees) for the 20 individual turns of the interferometer reported in Table 5.1. The average uncertainty of each determination is about ±20°. The band between the two horizontal lines corresponds to the global fit $\theta_0 = 24° ± 7°$. Each individual value could also be reversed by 180 degrees. The figure is taken from ref. [38].*

To understand the other aspect, concerning the consistency of data taken in different epochs of the year, let us first re-write the classical prediction for the fringe pattern at a given angle θ as

$$\left[\frac{\Delta\lambda(\theta;t)}{\lambda}\right]_{\text{class}} \sim \frac{D}{\lambda}\frac{v^2(t)}{c^2}\cos 2(\theta - \theta_0(t)) \tag{5.8}$$

where $v(t)$ and $\theta_0(t)$ indicate respectively the instantaneous magnitude and direction of the drift in the plane of the interferometer and contain the crucial information.

To find the relation with the earth motion, let us introduce a cosmic earth velocity with well defined magnitude V, right ascension α and angular declination γ. These parameters can be considered constant for short-time observations of a few days where there are no appreciable changes due to the earth orbital velocity around the sun. By restricting to this type of short-time observations, the only time dependence is then expected from the earth rotation and the standard identifications are $v(t) \equiv \tilde{v}(t)$ and $\theta_0(t) \equiv \tilde{\theta}_0(t)$ where $\tilde{v}(t)$ and $\tilde{\theta}_0(t)$ derive from the simple application of spherical trigonometry. Therefore, in this framework, one obtains [133]

$$\cos z(t) = \sin\gamma\sin\phi + \cos\gamma\cos\phi\cos(\tau - \alpha) \tag{5.9}$$

$$\frac{\tilde{v}_x(t)}{V} \equiv \sin z(t)\cos\tilde{\theta}_0(t) = \sin\gamma\cos\phi - \cos\gamma\sin\phi\cos(\tau - \alpha) \tag{5.10}$$

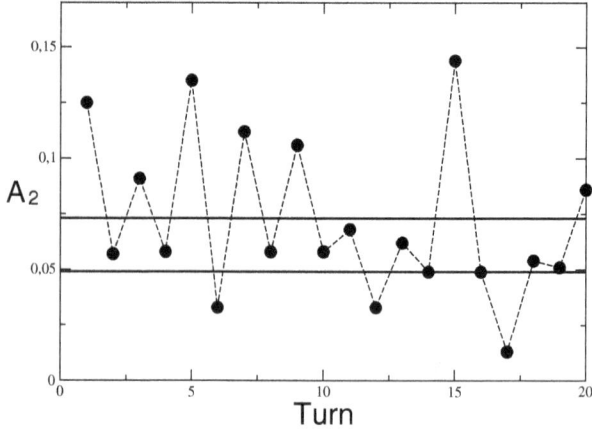

Fig. 5.2   *The 2nd-harmonic amplitude for the 20 individual turns of the interferometer reported in Table 5.1. The average uncertainty of each determination is about ±0.030. The band between the two horizontal lines corresponds to the global fit $A_2 = 0.061\pm0.012$. Within their errors, these individual values correspond to an observable velocity in the range $4\div14$ km/s. The figure is taken from ref. [38].*

$$\frac{\tilde{v}_y(t)}{V} \equiv \sin z(t) \sin \tilde{\theta}_0(t) = \cos\gamma \sin(\tau - \alpha) \qquad (5.11)$$

$$\tilde{v}(t) \equiv \sqrt{\tilde{v}_x^2(t) + \tilde{v}_y^2(t)} = V \sin z(t). \qquad (5.12)$$

Here $z = z(t)$ is the zenithal distance of $\mathbf{V}$, ϕ is the latitude of the laboratory, $\tau = \omega_{\text{sid}}t$ is the sidereal time of the observation in degrees ($\omega_{\text{sid}} \sim \frac{2\pi}{23^h 56'}$) and the angle θ_0 is counted conventionally from North through East so that North is $\theta_0 = 0$ and East is $\theta_0 = 90^o$.

Then, by re-writing the fringe shifts in the form

$$\left[\frac{\Delta\lambda(\theta; t)}{\lambda}\right]_{\text{class}} \sim 2C^{\text{class}}(t)\cos 2\theta + 2S^{\text{class}}(t)\sin 2\theta \qquad (5.13)$$

with

$$C^{\text{class}}(t) = \frac{D}{\lambda}\frac{v^2(t)}{2c^2}\cos 2\theta_0(t) \qquad\qquad S^{\text{class}}(t) = \frac{D}{\lambda}\frac{v^2(t)}{2c^2}\sin 2\theta_0(t) \qquad (5.14)$$

and assuming $v(t) = \tilde{v}(t)$ and $\theta_0(t) \equiv \tilde{\theta}_0(t)$ as in Eqs. (5.9)–(5.12), one arrives to the simple periodic structure

$$S^{\text{class}}(t) = S_{s1}^{\text{class}}\sin\tau + S_{c1}^{\text{class}}\cos\tau + S_{s2}^{\text{class}}\sin(2\tau) + S_{c2}^{\text{class}}\cos(2\tau) \quad (5.15)$$

$$C^{\text{class}}(t) = C_0^{\text{class}} + C_{s1}^{\text{class}} \sin\tau + C_{c1}^{\text{class}} \cos\tau + C_{s2}^{\text{class}} \sin(2\tau) + C_{c2}^{\text{class}} \cos(2\tau)$$
$$(5.16)$$

with Fourier coefficients C_k^{class} and S_k^{class} which are proportional to V^2 and depend on the latitude ϕ and on the angular parameters (α, γ).

Therefore, in this model, by fitting amplitude and phase of the 2nd-harmonic obtained from the fringe shifts, at different sidereal times, one can try to determine directly the cosmic parameters V, α and γ.

Miller's observations were made at Mt. Wilson in four epochs about April 1, August 1 and September 15, 1925, and later about February 8, 1926, with a total number of 6402 turns of the interferometer [7]. The result of his 1925 measurements, presented at the meeting in Kansas City on December 1925, was that "there is a positive, systematic ether-drift effect, corresponding to a relative motion of the earth and the ether, which at Mt. Wilson has an apparent velocity of 10 km/s" [7] and that the observed variations with the sidereal time "are exactly such as would be produced by a constant motion of the solar system in space toward an apex, near the north pole of the ecliptic having a right ascension of $17\frac{1}{2}$ hours and a declination of +65 degrees". At the same time, since the four sets of observations, performed during the year, gave no indication about the earth orbital motion, the velocity V was presumably much larger than 30 km/s. The observed, substantial reduction to 10 km/s, required by the size of the fringe shifts, was tentatively explained by Miller on the basis of Stokes' hypothesis of an ether which is partially entrained by matter. This might explain why the observed value of 10 km/s was only a small fraction of this much larger V.

Later on, in 1932, this picture of the earth cosmic motion was reconsidered by Miller with two significative changes. On the one hand, the apex was now pointing toward the south pole, namely with a right ascension of about 5 hours and declination of about −70 degrees. On the other hand, from tiny differences among the four different epochs, there was now some evidence for the earth orbital motion. By inspection of the yearly 'aberration circle', the cosmic component V was then estimated to be about 208 km/s. Again Stokes' partially-entrained-ether model was invoked by Miller to explain the strong suppression which was reducing this large kinematic velocity to the observed value of about 10 km/s.

Apart from this aspect, other authors, as for instance von Laue [130] and Thirring [131], objected to the overall consistency of this picture. Their argument, which was also re-proposed by Shankland et al. [47], depends on considering the daily averaged fringe shifts and amounts to the following. From relations (5.13), (5.15) and (5.16) one can construct the fringe pattern averaged over all sidereal times of the day. Then, by denoting $\langle ... \rangle$ the daily average of any given quantity, one finds, at any angle θ, the daily averaged fringe shift

$$\left\langle \left[\frac{\Delta\lambda(\theta)}{\lambda} \right] \right\rangle_{\text{class}} = 2C_0^{\text{class}} \cos 2\theta \qquad (5.17)$$

that can be cast into the form (C_0^{class}, which gives the daily average over τ, depends only on ϕ and γ) [47]

$$\left\langle \left[\frac{\Delta\lambda(\theta)}{\lambda} \right] \right\rangle_{\text{class}} = V^2 F(\gamma, \phi) \cos 2\theta. \qquad (5.18)$$

Therefore, since the latitude ϕ is a constant and the angular declination γ is fixed at any specific epoch, the daily averaged fringe shifts should all have a *common maximum* at $\theta = 0$. Only the amplitude can be different at different epochs. Instead, in Miller's observations the location of the maximum was differently displaced from the meridian (see Figs.25 of ref. [7] and Fig.3 of ref. [47]). This was a serious problem for the overall consistency of his observations.

We emphasize, however, that in this derivation any physical signal is assumed to produce the regular modulations expected from the earth rotation. As anticipated in Chapt.1, and discussed in Chapt.3 in connection with the Michelson-Morley experiment, one might be faced with a different scenario where the ether-drift is not a purely deterministic phenomenon. Then, the local instantaneous quantities $v_x(t) = v(t) \cos\theta_0(t)$ and $v_y(t) = v(t) \sin\theta_0(t)$ could differ non-trivially from those of the average earth motion $\tilde{v}_x(t)$ and $\tilde{v}_y(t)$, entering Eqs.(5.9)$-$(5.12), which just fix their typical boundaries. In this other perspective, combining observations of different days and different epochs becomes more delicate and there might be substantial deviations from Eq.(5.18). We shall return to this important point in Chapt.6.

For the moment, let us follow the historical development and consider other experiments that were performed just to check Miller's claims.

5.5 Kennedy-Illingworth 1926-1927

In response to Miller's announcements, a new ether-drift experiment was designed by Kennedy in 1926. As summarized in his contribution to the previously mentioned Conference on the Michelson-Morley experiments [102],

Table 5.2   *The infra-session averages $\langle D_A \rangle$ and $\langle D_B \rangle$ obtained from the 10 sets of rotations in each of the 32 sessions of Illingworth experiment. These values have been obtained from the weights of Illingworth Table III by applying the conversion factor 0.002.*

5 A.M. $\langle D_A \rangle$	5 A.M. $\langle D_B \rangle$	11 A.M. $\langle D_A \rangle$	11 A.M. $\langle D_B \rangle$	5 P.M. $\langle D_A \rangle$	5 P.M. $\langle D_B \rangle$	11 P.M. $\langle D_A \rangle$	11 P.M. $\langle D_B \rangle$
+0.00024	−0.00066	+0.00070	−0.00022	+0.00024	+0.00044	−0.00010	+0.00024
+0.00114	+0.00024	−0.00042	−0.00036	−0.00056	−0.00046	+0.00018	+0.00018
+0.00000	+0.00000	−0.00006	−0.00052	−0.00144	−0.00080	−0.00126	−0.00006
+0.00020	−0.00044	−0.00030	+0.00012	−0.00016	+0.00004	−0.00044	−0.00026
+0.00064	+0.00000	−0.00022	+0.00038	+0.00018	+0.00016	+0.00000	+0.00024
−0.00002	−0.00010	+0.00048	+0.00020	+0.00030	+0.00030	−0.00040	−0.00004
		−0.00014	−0.00006	+0.00030	+0.00014		
		−0.00006	+0.00004	+0.00036	−0.00036		
		−0.00006	+0.00016	+0.00006	−0.00006		
		+0.00000	+0.00024	−0.00010	+0.00010		

his small optical system, with a light path of 200 cm, was enclosed in an effectively insulated, sealed metal case containing helium at atmospheric pressure. Because of its small size, "...circulation and variation in density of the gas in the light paths were nearly eliminated. Furthermore, since the value of $\mathcal{N} - 1$ is only about 1/10 that for the air at the same pressure, the disturbing changes in density of the gas correspond to those in air to only 1/10 of the atmospheric pressure".

The essential ingredient of Kennedy's apparatus consisted in the introduction of a small step, 1/20 of wavelength thick, in one of the total reflecting mirrors of the interferometer allowing, in principle, for an ultimate fringe shift accuracy of about $1 \cdot 10^{-4}$. To reach this level of precision, Kennedy should have disposed of perfect mirrors and of a suitable (hotter) source of light. In the original version of the experiment, these refinements were not implemented giving an actual fringe shift accuracy of $2 \cdot 10^{-3}$. In these conditions, as Kennedy says explicitly [102], "...the velocity of 10 km/s found by Prof. Miller would produce a fringe shift corresponding to $8 \cdot 10^{-3}$", four times larger than the experimental resolution. Since the effect is quadratic in the velocity, Kennedy's result, fringe shifts $< 2 \cdot 10^{-3}$, can then be summarized as

$$v_{\text{obs}} < 5 \text{ km/s.} \tag{5.19}$$

Kennedy's apparatus was further refined by Illingworth in 1927 [21]. Besides improving the quality of the mirrors and of the source, Illingworth's data taking was also designed in order to account for the presence of steady thermal drift and of odd harmonics in the raw data. Looking at Illingworth's paper, one discovers that his refinements reached indeed the nominal $\mathcal{O}(10^{-4})$ accuracy mentioned by Kennedy, namely about $1/1500$ of wavelength for the individual readings and $(1 \div 2) \cdot 10^{-4}$ at the level of average values.

Let us then analyze Illingworth's results. He performed four series of observations in the first ten days of July 1927. These consisted of 32 experimental sessions, conducted daily at 5 A.M. (6), 11 A.M. (10), 5 P.M. (10) and 11 P.M.(6), in which he was measuring the fringe displacement caused by a rotation through a right angle of the apparatus. To take into account 90° rotations let us first re-write the fringe shifts as

$$\frac{\Delta\lambda(\theta)}{\lambda} = A_2 \cos 2(\theta - \theta_0). \tag{5.20}$$

Therefore Illingworth, in his first set (set A) of 10 rotations, North, East, South, West and back to North, was measuring $D_A \equiv 2A_2 \cos 2\theta_0$. In a second set (set B), North-East, North-West, South-West, South-East and back to North-East, performed immediately after the set A, he was then measuring $D_B \equiv 2A_2 \sin 2\theta_0$. Notice that both D_A and D_B differ from the positive-definite quantity $D \equiv 2A_2$ that should be inserted in Illingworth's numerical relation for his apparatus $v_{\text{obs}} = 112\sqrt{D}$. Therefore, the reported values for the two velocities $v_A = 112\sqrt{|D_A|}$ and $v_B = 112\sqrt{|D_B|}$ should only be taken as *lower* bounds for the true v_{obs}. The mean values $\langle D_A \rangle$ and $\langle D_B \rangle$ obtained from the 10 sets of rotations in the 32 individual sessions can be obtained from Illingworth's Table III and, for the convenience of the reader, are reported in our Table 5.2.

From Table 5.2, one finds that the quantity $\sqrt{\langle D_A \rangle^2 + \langle D_B \rangle^2}$ has a mean value of about 0.00045 ± 0.00035. If this is taken as an estimate of the positive definite D, it would correspond approximately to an observable velocity $v_{\text{obs}} \sim 2.4^{+0.8}_{-1.2}$ km/s. Equivalently, this observable velocity when compared to the classical prediction for 30 km/s $A_2^{\text{class}} \sim 0.035$ corresponds to an average 2nd-harmonic amplitude $\langle A_2^{\text{EXP}} \rangle \sim 0.00022 \pm 0.00017$.

However, this is only a partial view. To go deeper into Illingworth's experiment we have to consider his basic measurements, i.e. the individual turns of his interferometer. In this case, the only known basic set of data reported by Illingworth is set A of July 9th, 11 A.M. This set has been re-analyzed by Múnera [132] and his values for the fringe shifts are reported in our Table 5.3.

As one can see, the fringe shifts are not small and correspond to an observable velocity in the range 2-5 km/s. However, their sign seems to change randomly. Therefore, if one attempts to extract the observable

Table 5.3   *Illingworth's set A of July 9th, 11
A.M. as re-analyzed by Múnera [132].*

Rotation	D_A	$\lvert D_A \rvert$	v_A [km/s]
1	−0.00100	+0.00100	3.54
2	+0.00066	+0.00066	2.89
3	−0.00066	+0.00066	2.89
4	−0.00066	+0.00066	2.89
5	−0.00166	+0.00166	4.57
6	+0.00234	+0.00234	5.41
7	+0.00100	+0.00100	3.54
8	+0.00034	+0.00034	2.04
9	+0.00000	+0.00000	0.00
10	−0.00100	+0.00100	3.54

velocity from the mean of the 10 determinations, $\langle D_A \rangle \sim -0.00006$, the resulting value 0.9 km/s is much smaller than all individual determinations. The basis of Múnera's analysis was instead to estimate v_{obs} from $\langle \lvert D_A \rvert \rangle$ by obtaining a velocity $v_{\text{obs}} = 3.13 \pm 1.04$ km/s.

Now, the standard interpretation of the apparently random changes of sign is in terms of typical instrumental effects and the standard method for eliminating these is the original averaging procedure as employed by Illingworth. But, as already anticipated for the Michelson-Morley and Miller experiments, they could also indicate an unconventional form of stochastic drift which exhibits random fluctuations and which is only *indirectly* related to the macroscopic earth motion. In this alternative picture, Múnera's re-estimate would have a definite significance.

Clearly, by accepting this type of picture, further reduction of the data by performing inter-session averages ($\langle\langle ... \rangle\rangle$) can wash out completely the physical information contained in the original observations. In Table 5.4, we report the final inter-session averages $\langle\langle D_A \rangle\rangle$ and $\langle\langle D_B \rangle\rangle$ obtained by Illingworth for the various observation times.

Traditionally, from these final averages for $\langle\langle D_A \rangle\rangle$ at 5 A.M. and at 11 P.M. one has been deducing *upper bounds* $v_A \lesssim 2.12$ km/s and $v_A \lesssim 2.07$ km/s respectively. However, we have seen that these two estimates of v_A represent *lower* bounds for v_{obs}. Therefore, with an ether-drift of irregular nature the persistence of non-zero averages of 2 km/s implies that there were *many* values of v_{obs} which clearly had to be *larger* than both. This is consistent with Table 5.3 where, indeed, all values of v_A (except one) are larger than 2 km/s.

Table 5.4   *Illingworth's final inter-session averages.*

Observations	$\langle\langle D_A\rangle\rangle$	$\langle\langle D_B\rangle\rangle$
5 A.M.	$+0.00036 \pm 0.00012$	-0.00016 ± 0.00009
11 A.M.	-0.00001 ± 0.00007	-0.00000 ± 0.00006
5 P.M.	-0.00008 ± 0.00012	-0.00005 ± 0.00008
11 P.M.	-0.00034 ± 0.00014	$+0.00005 \pm 0.00006$

For this reason, in a different model of the drift, this 2 km/s velocity reported by Illingworth, rather than being interpreted as an upper bound would now represent a *lower* bound placed by his experiment. In this way, by combining with Kennedy's previous upper bound $v_{\text{obs}} < 5$ km/s, one would deduce that these two experiments, where light was propagating in helium at atmospheric pressure, provide a range for the observable velocity

$$\text{(Kennedy + Illingworth)} \qquad 2\,\text{km/s} \lesssim v_{\text{obs}} < 5\,\text{km/s} \tag{5.21}$$

in complete agreement with Múnera's determination $v_{\text{obs}} = (3.1 \pm 1.0)$ km/s.

In the following, however, to separate out the two experiments, we shall adopt the previous estimate

$$\text{Illingworth} \qquad v_{\text{obs}} \sim 2.4^{+0.8}_{-1.2}\,\text{km/s} \tag{5.22}$$

which was extracted from Table 5.2 by computing the average value of $\sqrt{\langle D_A\rangle^2 + \langle D_B\rangle^2}$. Then, by comparing with the classical prediction for 30 km/s $A_2^{\text{class}} \sim 0.035$ one finds a 2nd-harmonic amplitude $\langle A_2^{\text{EXP}}\rangle \sim 0.00022 \pm 0.00017$.

5.6   Piccard and Stahel 1926-1928

A small-size Michelson apparatus with photographic recording was realized by Piccard and Stahel [135–137]. In the years 1926–1928, they took measurements in a balloon (at heights of 2500 and 4500 m), on dry land in Brussels and on top of Mt.Rigi in Switzerland (at an height of 1800 m). Despite their optical path (280 cm) was much shorter than the size of the instruments used in the United States, they were convinced that the precision of their measurements was higher because spurious disturbances were less important.

Being aware that Miller's 1921 Mt. Wilson announcement of a non-zero ether-drift of about 9 km/s, if taken seriously, could undermine the basis of Einstein's relativity (Miller's results "carried a mortal blow to the theory of relativity"), they originally performed their measurements in a free balloon to check the dependence on altitude. In this first series of measurements thermal disturbances were so strong that they could only set an upper limit of about 9 km/s to the magnitude of any ether-drift. Meanwhile, in 1926 Miller had published [134] that the experiments in Cleveland, even those with Morley in 1905, once interpreted differently, were giving the same result as at Mt. Wilson, thus not confirming the original hypothesis of an increase with altitude. Therefore they decided to continue their measurements on ground at their laboratory in Brussels where thermal disturbances were much smaller and the precision became much higher.

As anticipated, their optical path had a length $D = 280$ cm. The mirrors were provided by the house of Jobin and Yvon and were chemically silver plated in their laboratory. The principal improvement with respect to the traditional direct observation was to register the fringe shifts by photographic recording. Finally, for thermal insulation, the interferometer was surrounded either by a thermostat filled with ice or by an iron enclosure where it could be possible to create a vacuum. This last solution was considered after having understood that the main instability in the fringe system was due to thermal disturbances in the air of the optical arms (rather than to temperature differences in the solid parts of the apparatus). However, very often the interference fringes were put out of order after few minutes due to the presence of residual bubbles of air in the vacuum chamber. For this reason, they finally decided to run the experiment at atmospheric pressure with the ice thermostat which, by its great heat capacity, was found to stabilize the temperature in a satisfactory way.

After the inconclusive result of the experiments in balloon, in their paper [137] Piccard and Stahel reported the individual measurements obtained in the other two environments: (i) laboratory in Brussels in November 1926 and (ii) Mt. Rigi in Switzerland in September 1927.

The measurements in Brussels were performed at midnight of November 25th and 29th and at 10:50 AM of November 23rd. Those at Mt. Rigi, on 16th and 17th September between 5 AM and 6 AM. From individual sets of ten rotations, they extracted amplitude and phase of the most probable 2nd harmonic which were giving respectively magnitude and direction of the local drift in the plane of the interferometer. As in Illingworth's experiment, the phase was found to vary in a completely arbitrary way. Therefore, by combining the data vectorially, the resultant velocity was much smaller than all individual determinations. The resulting observable velocity $v_{\text{obs}} \sim$ $(1.5 \div 1.7)$ km/s, which is traditionally reported for their experiment, was thus considered by Piccard and Stahel as a definite refutation of Miller's estimate $v_{\text{obs}} \sim 9$ km/s.

Again, however, if the phenomenon has an intrinsic non-deterministic nature, which induces random fluctuations in the direction of the local drift, such a vector average of the data will completely obscure the physical information contained in the original observations. Moreover, as discussed by Shankland et al. [47] (see also below), the profile of Miller's individual ether-drift data strongly differs from a standard Gaussian distribution. Therefore, deciding on the consistency between different experiments becomes more delicate. For this reason, a meaningful comparison with Miller can only be obtained by applying his same procedure to the Piccard-Stahel data. Namely, first to summarize each measurement into a definite pair of values $(A_2^{\text{EXP}}, \theta_0^{\text{EXP}})$ and then compute the average magnitude of the velocity from the measured amplitudes.

To this end, let us consider the values of the 2nd harmonic amplitude (corresponding each to sets of ten rotations) by starting from the midnight observations in Brussels. The individual values, in units 10^{-3}, were $A_2^{\text{EXP}} =$ 3.2, 5.2, 6.5, 2.2, 4.9, 3.8 from which one obtains the following mean and variance

$$\langle A_2^{\text{EXP}} \rangle = (4.3 \pm 1.5) \cdot 10^{-3} \qquad \text{(Brussels average night amplitude)}.$$

$$(5.23)$$

By inserting the values D=280 cm, and $\lambda = 0.4358 \cdot 10^{-4}$ cm and comparing with the relation[3]

$$A_2^{\text{EXP}} = \frac{D}{\lambda} \frac{v_{\text{obs}}^2}{c^2} \sim 0.064 \ \frac{v_{\text{obs}}^2}{(30 \text{ km/s})^2} \tag{5.24}$$

this gives an average observable velocity

$$v_{\text{obs}} \sim 7.8^{+1.2}_{-1.5} \text{ km/s} \tag{5.25}$$

If we repeat the same analysis for the Brussels morning data, from the individual amplitudes (again in units 10^{-3}) $A_2^{\text{EXP}} =$ 1.85, 1.27, 3.40, 1.00, 3.70, 1.14 one obtains

$$\langle A_2^{\text{EXP}} \rangle = (2.1 \pm 1.2) \cdot 10^{-3} \qquad \text{(Brussels average morning amplitude)}$$

$$(5.26)$$

---

[3] Notice that, in terms of their parameter A, Piccard and Stahel define the 2nd harmonic amplitude as $A_2 \equiv \frac{A}{2}$.

and an observable velocity

$$v_{\text{obs}} \sim 5.4^{+1.4}_{-1.8} \text{ km/s}. \tag{5.27}$$

Finally from the 12 Mt.Rigi observations, the measured amplitudes (always in units 10^{-3}) were $A_2^{\text{EXP}} = 3.4, 1.1, 4.0, 2.4, 2.4, 4.3, 2.3, 2.6, 0.6, 2.0, 1.2, 3.9$ so that one finds an average amplitude

$$\langle A_2^{\text{EXP}} \rangle = (2.5 \pm 1.2) \cdot 10^{-3} \qquad \text{(average Mt.Rigi amplitude)} \tag{5.28}$$

and an observable velocity

$$v_{\text{obs}} \sim 5.9^{+1.3}_{-1.6} \text{ km/s}. \tag{5.29}$$

By ignoring the small difference in the latitude of the laboratories (and also in the seasons of the year), from the above 24 amplitudes of the three sets of measurements, one can form global averages

$$\langle A_2^{\text{EXP}} \rangle = (2.8 \pm 1.5) \cdot 10^{-3} \qquad \text{(global average)} \tag{5.30}$$

and

$$v_{\text{obs}} \sim 6.3^{+1.5}_{-2.0} \text{ km/s} \qquad \text{(Piccard} - \text{Stahel)}. \tag{5.31}$$

To evaluate more precisely the probability content and thus to decide on the consistency with Miller's experiment, we have then computed the normalized probability histogram of the observable velocity from the above 24 individual amplitudes. The result is shown in panel (a) of Fig.5.3 and can be used to set a 75% Confidence Limit

$$4 \text{ km/s} \lesssim v_{\text{obs}} \lesssim 8 \text{ km/s} \qquad \text{(Piccard} - \text{Stahel 75\% C.L.)} \tag{5.32}$$

For Miller's experiment, we have instead extracted the observable velocity from the individual, original data in Figure 22d of ref. [7]. These data represent a rather large sample, representative of the whole Miller experiment, and the month of September (1925) is the closest to the months of September (1927) and November (1926) of the Piccard-Stahel measurements. The corresponding normalized probability histogram for Miller's experiment is reported in panel (c) of Fig.5.3.

Thus, by ignoring again the small difference in the latitude of the laboratories, we can estimate the compatibility of the two experiments from the overlap $S(P, Q)$ of the two histograms P and Q. This is shown in panel (b) of Fig.5.3 where the overlap area is calculated as [138]

$$S(P, Q) = \sum_{i=1}^{N} min(P_i, Q_i) = 0.645 \tag{5.33}$$

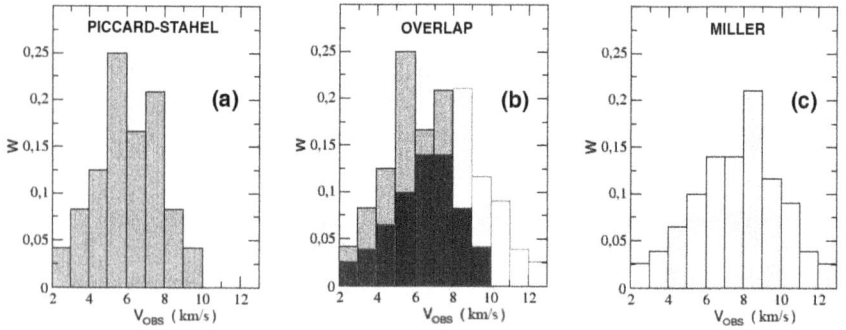

Fig. 5.3   *We report in panel (a) the probability histogram W for the observable velocity obtained through Eq.(5.24) from the 24 individual amplitudes reported by Piccard and Stahel. In panel (c) we report the analogous histogram obtained from Figure 22d of [7]. In both cases, the vertical normalization is to a unit area. Finally, in panel (b) we report the overlap of the two histograms. The area of the overlap is 0.645. This gives a consistency between the two experiments of about 64%.*

where P_i and Q_i are the bins of the two histograms in panels (a) and (c). On the other hand, the non-overlapping part is defined as

$$1 - S(P, Q) = 1/2 \sum_{i=1}^{N} |P_i - Q_i| = 0.355. \qquad (5.34)$$

This gives an estimate of the consistency of the two experiments of about 64% which is a quite high confidence level. At the same time, the fact that the agreement is restricted to the region $v_{obs} < 9$ km/s suggests that Miller's higher values are likely affected by systematic disturbances. This would confirm Piccard's and Stahel's claim that their apparatus, although of a smaller size, was more precise.

5.7   Michelson-Pease-Pearson 1926−1929

In early 1926, about forty years after the original 1887 ether-drift experiment, Michelson was busy with his improved measurements of the velocity of light between Mount Wilson and Mt. San Antonio [98]. However, after Miller's second announcement in December 1925, and especially after the publication of a confirming report [134] in April 1926, his claims could not be ignored. For this reason Walter Adams, the new director of Mt. Wilson Observatory, reminded Michelson that new improved ether-drift experiments were more important and "what the scientific world wants is *your* final word on the subject" [98].

Actually Michelson, together with his assistants Fred Pearson and Francis Pease, had already been designing and preparing a very large interferometer. Originally, it was to be made of low-expansion structural steel, of the same basic form of Miller's steel cross, with an effective light path of about fifty-five feet [98]. In June 1926 the apparatus was ready for operational tests. Although the results of this first series of trials were not advertised, the money spent and the experimental efforts indicate that Miller's claims were taken very seriously.

Michelson was not satisfied with this first apparatus. For this reason, in 1927-1928, further refinements were made to improve the temperature controls, the structural stability and increase the total light path up to eighty-five feet. After many measurements performed with this new apparatus, at a Meeting of the National Bureau of Standards by the Optical American Society in November 1928, Michelson reaffirmed the null ether-drift result for which he was most famous. His declaration at a press conference held the last day of the meeting was that "The results of my experiment conducted with greater scientific care, improved apparatus and refined technique, with the intention of eliminating every possible source of error, are again negative ... It is for physicists to study and explain these results and reconcile them with the existence of the hypothetical ether" [98].

Miller was also present and defended his experiment by saying that his measurements had been conducted "in the honest scope of arriving at a negative result too". He admitted that, in the end, his small positive effect could be due to periodic temperature differences in the air of the optical arms. However, on the basis of the actual experimental conditions, this was by no means clear.

After this brief historical introduction, let us now look in more detail at the Michelson, Pease and Pearson (MPP) results as reported in the two original papers [139, 140]. The authors do not report numbers so that we will quote literally from the two articles. In ref. [140] they say: "In the final series of experiments, the apparatus was transferred to a well-sheltered basement room of the Mount Wilson Laboratory. The length of the light path was increased to eighty-five feet, and the results showed that the precautions taken to eliminate temperature and pressure disturbances were effective. The results gave no displacement as great as *one-fiftieth* of that to be expected on the supposition of an effect due to a motion of the solar system of three hundred kilometers per second".

On the other hand, in ref. [139], after similar comments on the length of the apparatus and on the precautions taken to eliminate the various disturbances, one finds this other statement: "The results gave no displacement as great as *one-fifteenth* of that to be expected on the supposition of an effect due to a motion of the solar system of three hundred kilometers per second. These results are differences between the displacements observed at maximum and at minimum at sidereal times, the directions corresponding to Dr. Strömberg's calculations of the supposed velocity of the solar system". In the same paper, the authors report that, according to Strömberg's calculations " a displacement of 0.017 of the distance between fringes should have been observed at the proper sidereal times".

Clearly, although not explicitly stated, they were assuming some unknown mechanism which could largely reduce the fringe shifts with respect to the naive, huge non-relativistic value associated with a kinematical velocity of 300 km/s. Indeed, the classical expectation for the MPP apparatus were $A_2^{\text{class}} = \frac{L}{\lambda}\frac{(30\text{km/s})^2}{c^2} \sim 0.45$ for optical path of eighty-five feet and $A_2^{\text{class}} = \frac{L}{\lambda}\frac{(30\text{km/s})^2}{c^2} \sim 0.29$ for optical path of fifty-five feet.

To try to understand, we have looked at other sources. For instance, according to Shankland et al. [47] (see their Table I) the typical MPP fringe shifts were of the order of ±0.005. This estimate is likely taken from another paper which, rather surprisingly, was signed by Pease alone [141].

Pease declares that, in their experiment, to test Miller's claims, they concentrated on a purely *differential* type of measurement. For this reason, he only reports the difference

$$\delta(\theta) = \langle\frac{\Delta\lambda(\theta)}{\lambda}\rangle_{5.30} - \langle\frac{\Delta\lambda(\theta)}{\lambda}\rangle_{17.30} \tag{5.35}$$

between the mean fringe shifts $\langle\frac{\Delta\lambda}{\lambda}\rangle_{5.30}$, obtained after averaging over a large set of observations performed at sidereal time 5.30, and the mean fringe shifts $\langle\frac{\Delta\lambda}{\lambda}\rangle_{17.30}$ obtained after averaging in the same period at sidereal time 17.30. The quantity $\delta(\theta)$ has typical magnitude of ±0.004 or smaller.

As already anticipated, with a vector average of different observations one is assuming the conventional model of smooth modulations of the signal. Otherwise, if there were a substantial stochastic component in the signal, the cancelations introduced by a standard averaging process would become stronger and stronger by increasing the number of observations. Therefore, from these $\delta-$values, nothing can be said about the magnitude of the fringe shifts $\frac{\Delta\lambda(\theta)}{\lambda}$ obtained, before any averaging procedure and before any subtraction, in individual measurements at various hours of the day.

Pease just reports a poor-quality plot of a single observation, performed when the length of the optical path was still fifty-five feet, where the even fringe shift combinations Eq.(3.25) vary approximately in the range ±0.006. Its absolute value can then be taken to represent the magnitude of the 2nd-harmonic amplitude $A_2^{\text{EXP}} \sim 0.006$ for that observation (or equivalently $A_2^{\text{EXP}} \sim 0.009$ for the larger optical path of eighty-five feet). Its uncertainty, however, cannot be estimated without information on other individual sessions. Therefore, in this situation, we can only report an observable velocity with unknown error, namely

$$v_{\text{obs}} \sim 30 \text{ km/s} \sqrt{\frac{0.006 \pm ...}{0.29}} \sim (4.3 \pm ...) \text{ km/s.} \qquad (5.36)$$

We emphasize that Miller's extensive observations, as reported in Fig.22 of ref. [7] and in panel (c) of our histogram in Fig.5.3, show strong fluctuations with an observable velocity lying, within the errors, in the range $3-14$ km/s. This implies that, in individual measurements, the amplitude can change sometimes by the large factor $(14/3)^2 \sim 20$, thus supporting the picture of an ether-drift phenomenon of very irregular nature.

Thus for instance, for the MPP experiment, by comparing with the classical prediction for optical path of fifty-five feet $A_2^{\text{class}} \sim 0.29$, a very low observable velocity of 3 km/s implies that, in a single measurement, the experimental amplitude could become as small as $A_2^{\text{EXP}} \sim 0.003$. For this reason, it is fair to conclude that the only observation reported by Pease does not represent a refutation of Miller's experiment. This becomes even more true by noticing that this particular session, within a period of several months, was probably chosen to reinforce the idea of a negligibly small effect.

5.8 Joos 1930

One more classical ether-drift experiment was performed by Georg Joos in 1930 [22]. For the accuracy of the measurements (motor-driven rotation system, data collected during all 24 hours to cover the full sidereal day that were automatically recorded by photocamera), this experiment is by far the most precise one among the classical repetitions of the Michelson-Morley experiment.

The measured fringe shifts were typically of the order of one thousandth of wavelength which Joos decided to adopt as an upper limit placed by his experiment. Thus comparing with the classical relation for his apparatus

$$A_2^{\text{class}} = \frac{D}{\lambda} \frac{(30\text{km/s})^2}{c^2} \sim 0.375 \tag{5.37}$$

he deduced an observable velocity $v_{\text{obs}} \lesssim 1.5$ km/s.

We stress that Joos's optical system was enclosed in a hermetic housing and, traditionally, it has been always assumed that the fringe shifts were recorded in a partial vacuum. This is supported by several elements. For instance, when describing his device for electromagnetic fine movements of the mirrors, Joos explicitly refers to the condition of an evacuated apparatus, see p.393 of ref. [22]. This aspect is also confirmed by Miller who, quoting Joos' experiment, speaks of an "evacuated metal housing" in his 1933 article [7]. This is particularly important since later on, in 1934, Miller and Joos had a public letter exchange [52, 125] and Joos did not correct Miller's statement.

On the other hand, Swenson [98, 142] explicitly reports that fringe shifts were finally recorded by Joos with optical paths placed in a helium bath. Thus, in spite of the fact that this important aspect is never mentioned in Joos's papers, it seems more sound to follow Swenson's explicit statement.

In any case, whether his experiment was performed in vacuum or in a helium bath, Joos' results cannot be *directly* compared with those of Miller which were obtained in air at atmospheric pressure. This is because, as anticipated in Chapt.1, any possible non-zero light anisotropy is expected to become smaller and smaller when the refractive index of the gas tends to unity (as it happens by changing from air to gaseous helium and/or to vacuum). For this reason, beyond a standard classical analysis, there is no reason to consider Joos' measurements as a null result and/or a refutation of Miller's experiment. Our detailed re-analysis of his measurements in Chapt.6 will completely confirm this conclusion.

5.9   Criticism of Shankland's criticism of Miller's work

After the experiments performed by Kennedy, Illingworth, Piccard-Stahel, Michelson-Pease-Pearson and Joos, that apparently all disproved Miller's claims, a critical paper was written by Shankland and collaborators [47] in 1955, after Miller's death. In their reanalysis of his Mt. Wilson observations, they concluded that Miller's alleged ether-drift effect was actually due i) partly to statistical fluctuations and ii) partly to local temperature conditions. The Shankland team reanalysis is usually believed to have definitely closed the question.

To a closer look, however, their arguments are not so solid as they appear when reading the Abstract of their paper[4]. Indeed, within the paper these authors say that "...there can be little doubt that statistical fluctuations alone cannot account for the periodic fringe shifts observed by Miller" (see page 171 of ref. [47]). This is clearly illustrated by the histogram in their Figure 1 (reported by us as Fig.5.4) where the probability content deviates strongly from the typical gaussian shape expected for standard statistical fluctuations.

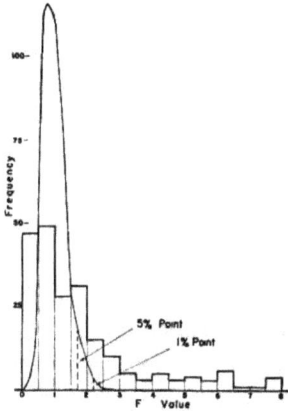

FIG. 1. The distribution of *F*-values for 216 sets of Mount Wilson data. The smooth theoretical curve is normalized so that the area under this curve is equal to the area of the histogram.

Fig. 5.4 *The probability histogram for 216 sets of Miller's observations as computed by Shankland et al. [47].*

Moreover, interpreting the observed effects on the basis of the local temperature conditions is certainly not the only possible explanation because "...we must admit that a direct and general quantitative correlation between amplitude and phase of the observed second harmonic on the one hand and the thermal conditions in the observation hut on the other hand could not be established" (see page 175 of ref. [47]).

The most surprising aspect of this interpretation in terms of *local* temperature conditions concerns, however, the extraction of Miller's average 2nd harmonic amplitude at Mt. Wilson (see Table III of ref. [47]). Indeed,

---

[4]A detailed rebuttal of the criticism raised by the Shankland team can be found in ref. [144].

their overall determination

$$A_2^{\mathrm{EXP}} = 0.044 \pm 0.022 \qquad (5.38)$$

when compared with the equivalent classical prediction for Miller's interferometer $A_2^{\mathrm{class}} = \frac{D}{\lambda} \frac{(30\mathrm{km/s})^2}{c^2} \sim 0.56$ corresponds to an average observable velocity

$$v_{\mathrm{obs}} \sim (8.4 \pm 2.2) \ \mathrm{km/s}. \qquad (5.39)$$

This is exactly the same observable velocity $v_{\mathrm{obs}} \sim 8.4$ km/s obtained from Miller's reanalysis of the Michelson-Morley experiment in Cleveland (see Miller's Figure 4 in our Chapt.3). Conceivably, their emphasis on the role of temperature effects would have been re-considered had they realized the perfect identity of two determinations obtained in completely different experimental conditions.

From this point of view, an interpretation in terms of temperature effects is only acceptable provided these effects have a *non-local* origin. For instance, within this alternative interpretation, and as anticipated in Chap.1, the ultimate explanation could depend on the temperature angular variations of about ±3 mK associated with the CMB dipole. Indeed, this would fit well with the typical magnitude of the temperature differences in the air of the optical arms of (1 ÷ 2) mK that were estimated by Kennedy, Shankland (see p. 175 of ref. [47], in particular footnote[16]) and Joos [52], to explain away Miller's observations. This basic non-locality of the effects in different experiments will be confirmed by our analysis of the temperature dependence of the gas refractive index in Chapt.7. As such, there are very good reasons to object to Shankland's criticism of Miller's work.

Chapter 6

What we experience as empty space is nothing but the configuration of the Higgs field that has the lowest possible energy. If we move from field jargon to particle jargon, this means that empty space is actually filled with Higgs particles. They have Bose condensed.

G. 't HOOFT, In Search of the Ultimate Building Blocks, 1997.

6.1 The non-trivial vacuum of present particle physics

Today, before addressing the outcome and the interpretation of the ether-drift experiments, one should take into account two basic ingredients: (i) the representation of the vacuum in present particle physics and (ii) the discovery of an anisotropy of the Cosmic Microwave Background (CMB). In this section we will consider point (i) and address point (ii) in the following section.

The discovery of the Higgs boson at LHC has confirmed the basic idea of spontaneous symmetry breaking where particle masses originate from the particular structure of the vacuum. This should not be thought as trivially empty but, instead, as a "condensate" of the Higgs field quanta. These condense because their trivially empty vacuum is a meta-stable state and not the lowest energy state of the theory.

This situation can be summarized by saying [54] that "What we experience as empty space is nothing but the configuration of the Higgs field that has the lowest possible energy. If we move from field jargon to particle jargon, this means that empty space is actually filled with Higgs particles. They have Bose condensed". The explicit translation from field jargon to particle jargon, with the substantial equivalence between the effective potential of quantum field theory and the energy density of a dilute particle condensate, can be found for instance in ref. [55].

The symmetric vacuum, with a vanishing Higgs field, will eventually be re-established by heating the system above a critical temperature $T = T_c$ where the condensate "evaporates". This temperature in the Standard Model is so high that one usually approximates the ordinary vacuum as a zero-temperature system (think of ^{4}He at a temperature 10^{-12} K). This explains why the physical vacuum might behave as a form of superfluid medium [145] which is not trivially empty but through which bodies flow without any apparent friction. To better appreciate this remark, compare with the analogous case of liquid helium in Fig.6.1.

Clearly, this form of quantum vacuum is not the kind of ether imagined by Lorentz. However, if possible, this modern view of the vacuum state is even more different from the empty space-time of Special Relativity that Einstein had in mind in 1905. Therefore, one may ask if Bose condensation, i.e. the macroscopic occupation of the same quantum state, say $\mathbf{k} = 0$ in some reference frame Σ, can represent the operative construction of a "quantum ether" [146]. This characterizes the *physically realized form of relativity* and could play the role of preferred reference frame in a modern Lorentzian approach.

This possibility is usually not considered with the motivation, perhaps, that the average properties of the condensed phase are summarized into a single quantity which transforms as a world scalar under the Lorentz group, for instance, in the Standard Model, the vacuum expectation value $\langle \Phi \rangle$ of the Higgs field. However, this does not necessarily imply that the vacuum state itself has to be *Lorentz invariant*. Namely, Lorentz transformation operators U', U'',...could transform non trivially the reference vacuum state[1] $|\Psi^{(0)}\rangle$ (appropriate to an observer at rest in Σ) into $|\Psi'\rangle$, $|\Psi''\rangle$,... (appropriate to moving observers S', S'',..) and yet, for any Lorentz-invariant operator G, one would find

$$\langle G \rangle_{\Psi^{(0)}} = \langle G \rangle_{\Psi'} = \langle G \rangle_{\Psi''} = .. \tag{6.1}$$

---

[1] We ignore here the problem of vacuum degeneracy by assuming that any overlapping among equivalent vacua vanishes in the infinite-volume limit of quantum field theory (see e.g. S. Weinberg, *The Quantum Theory of Fields*, Cambridge University Press, Vol.II, pp. 163-167).

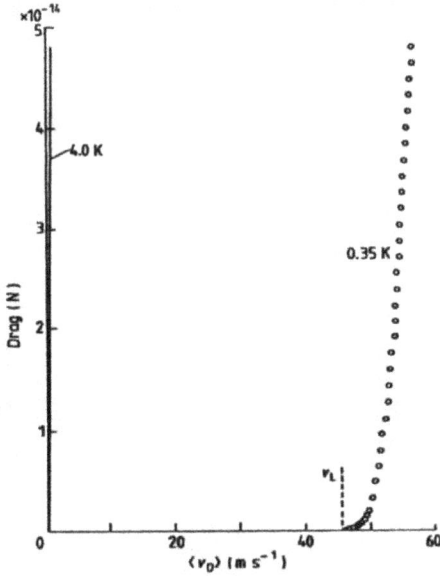

**Figure 2.9 Drag on a negative ion moving through liquid He II at 0.35 K and 25 atm,
as a function of the average ionic drift velocity $\langle v_D \rangle$. Drag sets in when $\langle v_D \rangle$ reaches
critical value which is close to roton v_L. In contrast, drag on ion in liquid He I at 4.0 K
sets in as soon as it starts to move, showing that superfluidity is absent. (After Allum et
al 1977.)**

Fig. 6.1   *The drag force on negative ions drifting through liquid ^{4}He. Measurements
were performed at two temperatures above and below the λ−point. The figure is taken
from 'Superfluidity and Superconductivity', by D.R. Tilley and J. Tilley, A. Hilger Ltd,
1986.*

Now, according to general quantum field theoretical arguments (see e.g.
[56]), deciding on the Lorentz invariance of the vacuum requires to consider
the eigenvalues and the algebra of the *global* Poincaré operators P_α, $M_{\alpha,\beta}$
(α ,β=0, 1, 2, 3) where P_α are the 4 generators of the space-time transla-
tions and $M_{\alpha\beta} = -M_{\beta\alpha}$ are the 6 generators of the Lorentzian rotations
with commutation relations

$$[P_\alpha, P_\beta] = 0 \tag{6.2}$$

$$[M_{\alpha\beta}, P_\gamma] = \eta_{\beta\gamma} P_\alpha - \eta_{\alpha\gamma} P_\beta \tag{6.3}$$

$$[M_{\alpha\beta}, M_{\gamma\delta}] = \eta_{\alpha\gamma} M_{\beta\delta} + \eta_{\beta\delta} M_{\alpha\gamma} - \eta_{\beta\gamma} M_{\alpha\delta} - \eta_{\alpha\delta} M_{\beta\gamma} \tag{6.4}$$

where $\eta_{\alpha\beta} = \mathrm{diag}(1, -1, -1, -1)$ is the Minkowski tensor. In this framework, as discussed in refs. [38, 147, 148], exact Lorentz invariance of the

vacuum requires to impose the problematic condition of a vanishing vacuum energy. As an example, one can consider the generator of a Lorentz-transformation along the 1-axis M_{01} for which one finds

$$P_1 M_{01} |\Psi^{(0)}\rangle = M_{01} P_1 |\Psi^{(0)}\rangle + P_0 |\Psi^{(0)}\rangle. \tag{6.5}$$

Therefore, even assuming zero spatial momentum for the vacuum condensation phenomenon, a non zero vacuum energy E_0 implies

$$P_1 M_{01} |\Psi^{(0)}\rangle = E_0 |\Psi^{(0)}\rangle \neq 0. \tag{6.6}$$

This means that the state $M_{01} |\Psi^{(0)}\rangle$ is non vanishing so that the reference vacuum state $|\Psi^{(0)}\rangle$ cannot be Lorentz invariant.

The simplest consequence of such non-invariance of the vacuum is an energy-momentum flow along the direction of motion with respect to Σ. In fact, by defining a boosted vacuum state $|\Psi'\rangle$ as

$$|\Psi'\rangle = e^{\lambda' M_{01}} |\Psi^{(0)}\rangle \tag{6.7}$$

(recall that $M_{01} \equiv -iL_1$ is an anti-hermitian operator) and using the relations

$$e^{-\lambda' M_{01}} P_1 e^{\lambda' M_{01}} = \cosh \lambda' \, P_1 + \sinh \lambda' \, P_0 \tag{6.8}$$

$$e^{-\lambda' M_{01}} P_0 e^{\lambda' M_{01}} = \sinh \lambda' \, P_1 + \cosh \lambda' \, P_0 \tag{6.9}$$

one finds

$$\langle P_1 \rangle_{\Psi'} = E_0 \sinh \lambda' \qquad\qquad \langle P_0 \rangle_{\Psi'} = E_0 \cosh \lambda'. \tag{6.10}$$

Clearly this result contrasts with the alternative approach where one tends to consider E_0 as a spurious concept and rather tries to characterize the vacuum through a *local* energy-momentum tensor of the form [149, 150]

$$\langle W_{\mu\nu} \rangle_{\Psi^{(0)}} = \rho_v \, \eta_{\mu\nu} \tag{6.11}$$

(ρ_v being a space-time independent constant). In this case, one is driven to completely different conclusions. In fact, by introducing the Lorentz transformation matrices Λ^μ_ν to any moving frame S', defining $\langle W_{\mu\nu} \rangle_{\Psi'}$ through the relation

$$\langle W_{\mu\nu} \rangle_{\Psi'} = \Lambda^\sigma_{\ \mu} \Lambda^\rho_{\ \nu} \langle W_{\sigma\rho} \rangle_{\Psi^{(0)}} \tag{6.12}$$

and using Eq.(6.11), the expectation value of W_{0i} in any boosted vacuum state $|\Psi'\rangle$ vanishes, just as it vanishes in $|\Psi^{(0)}\rangle$, so that

$$\int d^3x \, \langle W_{0i} \rangle_{\Psi'} \equiv \langle P_i \rangle_{\Psi'} = 0 \tag{6.13}$$

Still, the idea to simply get rid of E_0 gives rise to some problems. For instance, in a second-quantized formalism, single-particle energies $E_1(\mathbf{p})$ are defined as the energies of the corresponding one-particle states $|\mathbf{p}\rangle$ minus the energy of the zero-particle, vacuum state. If E_0 is considered a spurious concept, $E_1(\mathbf{p})$ will also become an ill-defined quantity. At the same time, the idea to characterize the physical vacuum through its energy E_0 has solid motivations. The ground state, in fact, is by definition the state with lowest energy as obtained from the solution of a minimum problem. As such, it should correspond to an energy eigenstate in view of the standard equivalence between eigenvalue equation and Rayleigh-Ritz variational procedure.

Finally, at a deeper level, one should also realize that in an approach based solely on Eq.(6.11) the properties of $|\Psi^{(0)}\rangle$ under a Lorentz transformation are not well defined. In fact, a transformed vacuum state $|\Psi'\rangle$ is obtained, for instance, by acting on $|\Psi^{(0)}\rangle$ with the boost generator M_{01}. Once $|\Psi^{(0)}\rangle$ is considered an eigenstate of the energy-momentum operator, one can definitely show that, for $E_0 \neq 0$, $|\Psi'\rangle$ and $|\Psi^{(0)}\rangle$ differ nontrivially. On the other hand, if $E_0 = 0$ there are only two alternatives: either $M_{01}|\Psi^{(0)}\rangle = 0$, so that $|\Psi'\rangle = |\Psi^{(0)}\rangle$, or $M_{01}|\Psi^{(0)}\rangle$ is a state vector proportional to $|\Psi^{(0)}\rangle$, so that $|\Psi'\rangle$ and $|\Psi^{(0)}\rangle$ differ by a phase factor. Therefore, if the structure in Eq.(6.11) were really equivalent to impose the exact Lorentz invariance of the vacuum, it should be possible to show similar results, for instance that such a $|\Psi^{(0)}\rangle$ state can remain invariant under a boost, i.e. be an eigenstate of

$$M_{0i} = -i \int d^3x \, (x_i W_{00} - x_0 W_{0i}) \tag{6.14}$$

with zero eigenvalue. However, there is no way to obtain such a result by just starting from Eq.(6.11) (this only amounts to the weaker condition $\langle M_{0i}\rangle_{\Psi^{(0)}} = 0$). Thus, it should not come as a surprise that one can run into contradictory statements and it is not obvious that the local relations (6.11) represent a more fundamental approach to the vacuum.

While a non-zero vacuum energy $E_0 \neq 0$ might have different explanations, one can try to explore the possible consequences of its existence by just observing that, in interacting quantum field theories, there is no known way to ensure consistently the condition $E_0 = 0$ without imposing an *unbroken supersymmetry* (which is not phenomenologically acceptable). This makes the issue of an exact Lorentz invariant vacuum a difficult problem which, at present, cannot be solved on purely theoretical grounds[2].

---

[2]One could also argue that a satisfactory solution of the vacuum energy problem lies

Nevertheless, to decide whether the laboratory frame S' defines (or not) a true state of rest, one could address the problem experimentally and try to observe the hypothetical energy-momentum flow. Here is where one gets in touch with ether-drift experiments. Indeed, optical interferometry represents a powerful tool to detect unexpected modifications in the structure of matter by studying light propagation inside the various media. Within the analogy of the vacuum with superfluid helium, this would have a definite counterpart. In fact, strictly speaking, superfluid helium still contains an infinitesimal fraction of 'normal' fluid which only vanishes at $T = 0$. The residual drag induced by this normal component [156], although smaller by many orders of magnitude than the corresponding drag force above the λ−point (and thus not appreciable on the scale of Fig.6.1), remains then non-zero.

6.2   The Cosmic Microwave Background

Precise observations with aircrafts and satellites have revealed a tiny anisotropy in the temperature of the Cosmic Microwave Background (CMB) [50, 51]. The present interpretation of its dominant dipole component (the *kinematic* dipole [49]) is in terms of a Doppler effect ($\beta = v/c$)

$$T(\theta) = \frac{T_o\sqrt{1 - \beta^2}}{1 - \beta\cos\theta} \tag{6.15}$$

due to a motion of the solar system with average velocity $v \sim 370$ km/s toward a point in the sky of right ascension $\alpha \sim 168°$ and declination

---

definitely beyond flat space. Nevertheless, in the absence of a consistent quantum theory of gravity, physical models of the vacuum in flat space can be useful to clarify a crucial point that, so far, remains obscure: the huge difference which is seen when comparing the typical vacuum-energy scales of particle physics with the value of the cosmological term needed in Einstein's equations to fit the observations. In fact, the picture of the vacuum as a superfluid can naturally explain why there might be no non-trivial macroscopic curvature in the equilibrium state where any liquid is self-sustaining [145]. In any liquid, in fact, curvature requires *deviations* from the equilibrium state. In such representation of the lowest energy state, where the arbitrarily large condensation energy of the liquid plays no observable role, one can intuitively understand why curvature effects can be orders of magnitude smaller than those naively expected. In this perspective, 'emergent-gravity' approaches [151–153], where gravity somehow arises from long-wavelength excitations of the same physical flat-space vacuum, may become natural and, to find the appropriate infinitesimal value of the cosmological term [154, 155], we are lead to sharpen our understanding of the vacuum structure and of its excitation mechanisms by starting from the picture of a superfluid medium in flat space. This type of picture will be explored in Chap.7.

$\gamma \sim -7°$. Therefore, if one sets $T_o \sim 2.7$ K and $\beta \sim 0.0012$, as for $v \sim 370$ km/s, there are angular variations of a few millikelvin

$$\Delta T(\theta) \sim T_o \beta \cos \theta \sim \pm 3 \text{ mK}. \tag{6.16}$$

By accepting this interpretation, a first question naturally arises: could one detect these tiny variations with measurements performed entirely within the earth laboratory? After all, the CMB is an all-pervading medium and aircrafts and satellites are nothing but aerials carried along the same cosmic motion of the earth. They are certainly more sensitive than aerials placed on the roof of a building or inside a laboratory but the principle is exactly the same.

Now, if nothing prevents in principle to detect the CMB dipole with measurements performed inside the earth laboratory, we may proceed one step further and ask a second question: could one detect the same earth motion by observing an "ether drift", i.e. a small difference of the velocity of light propagating in different directions?

To understand this point, let us consider the two-way velocity of light $\bar{c}_\gamma(\theta)$ and its anisotropy $\Delta \bar{c}_\theta = \bar{c}_\gamma(\pi/2 + \theta) - \bar{c}_\gamma(\theta)$. If we introduce a direction-dependent refractive index through the relation

$$\bar{c}_\gamma(\theta) \equiv \frac{c}{\mathcal{N}(\theta)} \tag{6.17}$$

for gaseous systems, where $\bar{\mathcal{N}}(\theta) - 1 \ll 1$, one finds

$$\frac{\Delta \bar{c}_\theta}{c} \sim \bar{\mathcal{N}}(\theta) - \bar{\mathcal{N}}(\pi/2 + \theta). \tag{6.18}$$

Now, at room temperature the air refractive index changes typically as $(\partial \mathcal{N}_{\text{air}}/\partial T) \sim 10^{-6}$ K^{-1}. Therefore it is not inconceivable that the light anisotropy observed in the classical experiments, namely $\frac{\Delta \bar{c}_\theta}{c} = \mathcal{O}(10^{-10})$, could indeed arise from angular temperature differences of the order of one millikelvin. After all, temperature differences of this typical magnitude, which could induce convection of the gas molecules and density variations in the air of the optical arms, were considered by Kennedy, Shankland and Joos [47, 52], as a possible mechanism to explain away Miller's alleged ether-drift effect.

Given this premise, let us explore in some more detail this idea of convection currents of the gas molecules and consider light propagation in a gaseous system of refractive index $\mathcal{N}$. By assuming isotropy, the time t spent by refracted light to cover some given distance L within the medium is $t = \mathcal{N}L/c$. This can be expressed as the sum of $t_0 = L/c$ and

$t_1 = (\mathcal{N} - 1)L/c$ where t_0 is the same time as in the vacuum and t_1 represents the additional, average time by which refracted light is slowed down by the presence of matter. If there are convective currents, due to the motion of the laboratory with respect to a preferred reference frame Σ, then t_1 will be different in different θ directions and there will be an anisotropy of the velocity of light proportional to $(\mathcal{N} - 1)$.

To make this more explicit, let us consider light propagating in a 2-dimensional plane and express t_1 as

$$t_1 = \frac{L}{c} f(\mathcal{N}, \theta, \beta) \tag{6.19}$$

with $\beta = v/c$, v being (the projection onto the considered plane of) the relevant velocity with respect to Σ where the isotropic form

$$f(\mathcal{N}, \theta, 0) = \mathcal{N} - 1 \tag{6.20}$$

is assumed. By expanding around $\mathcal{N} = 1$ where, whatever β, f vanishes by definition, one finds for gaseous systems (where $\mathcal{N} - 1 = \epsilon \ll 1$) the universal trend

$$f(\mathcal{N}, \theta, \beta) \sim \epsilon\, F(\theta, \beta) \tag{6.21}$$

with

$$F(\theta, \beta) \equiv (\partial f / \partial \mathcal{N})|_{\mathcal{N}=1} \tag{6.22}$$

and $F(\theta, 0) = 1$. Therefore, by introducing the one-way velocity of light

$$t(\mathcal{N}, \theta, \beta) = \frac{L}{c_\gamma(\mathcal{N}, \theta, \beta)} \sim \frac{L}{c} + \frac{L}{c} \epsilon\, F(\theta, \beta) \tag{6.23}$$

one gets

$$c_\gamma(\mathcal{N}, \theta, \beta) \sim \frac{c}{\mathcal{N}} \left[1 - \epsilon\, (F(\theta, \beta) - 1) \right] \tag{6.24}$$

Analogous relations hold for the two-way velocity $\bar{c}_\gamma(\mathcal{N}, \theta, \beta)$

$$\bar{c}_\gamma(\mathcal{N}, \theta, \beta) = \frac{2\, c_\gamma(\mathcal{N}, \theta, \beta) c_\gamma(\mathcal{N}, \pi + \theta, \beta)}{c_\gamma(\mathcal{N}, \theta, \beta) + c_\gamma(\mathcal{N}, \pi + \theta, \beta)}$$

$$\sim \frac{c}{\mathcal{N}} \left[1 - \epsilon\, \left(\frac{F(\theta, \beta) + F(\pi + \theta, \beta)}{2} - 1 \right) \right]. \tag{6.25}$$

A more explicit expression can be obtained by exploring some general properties of the function $F(\theta, \beta)$. By expanding in powers of β

$$F(\theta, \beta) - 1 = \beta F_1(\theta) + \beta^2 F_2(\theta) + \dots \tag{6.26}$$

and taking into account that, by the very definition of two-way velocity, $\bar{c}_\gamma(\mathcal{N}, \theta, \beta) = \bar{c}_\gamma(\mathcal{N}, \theta, -\beta)$, it follows that $F_1(\theta) = -F_1(\pi + \theta)$. Therefore, to $\mathcal{O}(\beta^2)$, by expressing the combination $F_2(\theta) + F_2(\pi + \theta)$ as an infinite expansion of even-order Legendre polynomials with arbitrary coefficients, we finally get the general structure

$$\bar{c}_\gamma(\mathcal{N}, \theta, \beta) \sim \frac{c}{\mathcal{N}} \left[1 - \epsilon\,\beta^2 \sum_{n=0}^{\infty} \zeta_{2n} P_{2n}(\cos\theta) \right] \qquad (6.27)$$

Eq.(6.27) implies a fractional anisotropy $\frac{\Delta\bar{c}_\theta}{c} \sim \epsilon\beta^2$. Therefore, by replacing the values $\beta \sim 0.0012$, and $\epsilon \sim 2.8 \cdot 10^{-4}$ or $\epsilon \sim 3.3 \cdot 10^{-5}$, respectively for air or gaseous helium at room temperature and atmospheric pressure, we obtain the right order of magnitude, $\frac{\Delta\bar{c}_\theta}{c} = \mathcal{O}(10^{-10})$ and $\frac{\Delta\bar{c}_\theta}{c} = \mathcal{O}(10^{-11})$, found in the classical experiments (Michelson-Morley, Miller, Kennedy, Illingworth...). The surprising conclusion is that those original measurements could indeed represent the first experimental indication for the earth motion within the CMB.

Truly enough, the CMB is a definite medium with a rest frame where its dipole anisotropy is zero. Motion with respect to this frame can be detected and, in fact, has been detected. In this sense, the idea that small modifications of the gaseous matter, produced by tiny temperature variations, can be detected in a laboratory by precise optical measurements, while certainly unconventional, would not have the revolutionary implications of a genuine preferred-frame effect due to the vacuum structure.

Still, it should not be overlooked that the observed CMB dipole can be reconstructed, to good approximation, by combining the various peculiar motions which are involved, namely the rotation of the solar system around the galactic center, the motion of the Milky Way around the center of the Local Group and the motion of the Local Group of galaxies in the direction of that large concentration of matter known as the Great Attractor [51]. In this way, once a vanishing CMB dipole is equivalent to switching-off all possible peculiar motions, one naturally arrives to the concept of a global frame of rest which is fixed by the average distribution of matter in the universe.

But such global frame could also reflect a vacuum structure which, as discussed in the previous section, has some degree of substantiality and, thus, could characterize non-trivially the form of relativity which is physically realized in nature. In this case, the isotropy of the CMB radiation might just *indicate* the existence of such a global frame that we could decide to call the "ether", but the CMB itself would *not* coincide with this type of ether.

To explore this more radical alternative, one can start to look at experiments where optical cavities are maintained in an extremely high vacuum, both at room temperature and in the cryogenic regime. The reason is that, in this limit, where any residual gaseous matter is totally negligible, a tiny temperature gradient of a fraction of millikelvin will not produce any observable light anisotropy. As such, the persistence of some infinitesimal effect in the vacuum limit would definitely support the idea of a genuine preferred frame for relativity. Therefore ether-drift measurements performed in different experimental conditions (e.g. in gaseous media vs. vacuum) can provide precious complementary indications. These more elaborated aspects will be discussed in Chap.7.

6.3   General aspects of the ether-drift experiments

As anticipated, nowadays, the original 1887 Michelson-Morley experiment [1] and its early repetitions performed at the turn of 19th and 20th centuries (by Miller [7], Kennedy [102], Illingworth [21], Joos [22] ...) are usually considered as being well understood. Instead, all emphasis is on their modern versions with lasers stabilized by optical cavities, see e.g. [23] for a review. These modern experiments adopt a very different technology but, in the end, have exactly the same scope: searching for the possible existence of a preferred reference frame through an anisotropy of the two-way velocity of light $\bar{c}_\gamma(\theta)$.

As discussed in the two previous sections, today, there might be good reasons to look for a non-trivial angular dependence of $\bar{c}_\gamma(\theta)$. While these motivations are very different from those of the classical theory, in the end, the theoretical predictions can be placed in a form which mimics the classical calculation. Therefore, to clearly display differences and analogies, it is better to re-propose the whole analysis from scratch.

In terms of the anisotropy $\Delta\bar{c}_\theta = \bar{c}_\gamma(\pi/2 + \theta) - \bar{c}_\gamma(\theta)$ the time difference $\Delta t(\theta)$ for light propagation back and forth along perpendicular rods of length D can be expressed as

$$\Delta t(\theta) = \frac{2D}{\bar{c}_\gamma(\theta)} - \frac{2D}{\bar{c}_\gamma(\pi/2 + \theta)} \sim \frac{2D}{c}\frac{\Delta\bar{c}_\theta}{c} \qquad (6.28)$$

(where, in the last relation, we have assumed that light propagates in a medium of refractive index $\mathcal{N} = 1 + \epsilon$, with $\epsilon \ll 1$). We have also assumed the validity of Lorentz transformations so that the length of a rod does not depend on its orientation, in the frame S' where it is at rest[3]. In this way,

---

[3]Regardless of Lorentz transformations, a fundamental anisotropy of lengths in the

one gets the fringe patterns (λ is the light wavelength)

$$\frac{\Delta\lambda(\theta)}{\lambda} \sim \frac{2D}{\lambda} \frac{\Delta\bar{c}_\theta}{c} \qquad (6.29)$$

which were measured with Michelson interferometers. Notice that, at this stage, this is formally the same relation (1.4) as in the classical theory. The only difference concerns the anisotropy. We know that the classical prediction was $\frac{\Delta\bar{c}_\theta}{c} \sim \beta^2/2$. Instead, by assuming the validity of Lorentz transformations, one expects a very different value.

To be definite, let us assume that the medium fills an optical cavity at rest in the laboratory frame S' which moves with uniform velocity v with respect to the hypothetical preferred frame Σ. If we assume i) that the velocity of light is exactly isotropic when $S' \equiv \Sigma$ and ii) the validity of Lorentz transformations, it follows that any anisotropy in S' should vanish identically either for $v = 0$ or for $\mathcal{N} = 1$, i.e. when the velocity of light c_γ coincides with the basic parameter c entering Lorentz transformations. Thus, one can expand in powers of the two small parameters ϵ and $\beta = v/c$. By taking into account that, by its very definition, the two-way velocity $\bar{c}_\gamma(\theta)$ is invariant under the replacement $\beta \to -\beta$ and that, for any fixed β, is also invariant under the replacement $\theta \to \pi + \theta$, to lowest non-trivial level $\mathcal{O}(\epsilon\beta^2)$, one finds the general expression [38, 40]

$$\bar{c}_\gamma(\theta) \sim \frac{c}{\mathcal{N}} \left[1 - \epsilon\, \beta^2 \sum_{n=0}^{\infty} \zeta_{2n} P_{2n}(\cos\theta)\right]. \qquad (6.30)$$

Here, to take into account invariance under $\theta \to \pi + \theta$, the angular dependence has been given as an infinite expansion of even-order Legendre polynomials with arbitrary coefficients $\zeta_{2n} = \mathcal{O}(1)$. In Einstein's special relativity, where there is no preferred reference frame, these ζ_{2n} coefficients should vanish identically. In a "Lorentzian" approach, on the other hand, there is no reason why they should vanish *a priori*.

Notice that we have found the same general structure as in Eq.(6.27) which was deduced by assuming the existence of convective currents of

---

laboratory frame (i.e. not due to mechanical deformations under rotation) would amount to an anisotropy of the underlying molecular forces and, as such, of the basic atomic parameters. This possibility is severely limited experimentally. In fact, in the most recent versions of the original Hughes-Drever experiment [157, 158], where one measures the atomic energy levels as a function of their orientation with respect to the fixed stars, possible deviations from isotropy have been found below the 10^{-20} level [159]. This is incomparably smaller than any other effect on the velocity of light we are going to discuss. Therefore an angular dependence of lengths, if present, is completely negligible and, from now on, we will assume $D(\theta) = D =$ constant.

the gas molecules associated with the motion of the laboratory. In this way, as anticipated in the Introduction, the same basic result $\frac{\Delta \bar{c}_\theta}{c} \sim \epsilon \beta^2$ originates from simple symmetry arguments but find additional justification in underlying physical mechanisms. We also emphasize that the structure $\frac{\Delta \bar{c}_\theta}{c} \sim \epsilon \beta^2$ has been deduced for light propagation in dielectric media whose refractive index is infinitesimally close to unity. As such, it cannot be extended to solid dielectrics.

Concerning the modern experiments, a possible anisotropy of $\bar{c}_\gamma(\theta)$ would show up through the relative frequency shift, i.e. the beat signal, $\Delta \nu(\theta)$ of two orthogonal optical resonators. Their frequency

$$\nu(\theta) = \frac{\bar{c}_\gamma(\theta) m}{2L} \tag{6.31}$$

is proportional to the two-way velocity of light within the resonator through an integer number m, which fixes the cavity mode, and the length of the cavity L as measured in the laboratory frame S', see Fig.6.2.

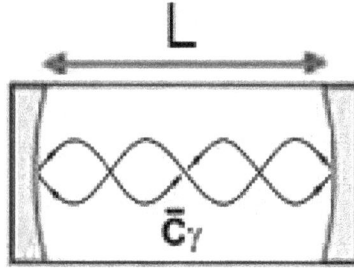

Fig. 6.2   *The idealized scheme of a modern optical cavity.*

Therefore, once the length of a cavity in its rest frame does not depend on its orientation, one finds

$$\frac{\Delta \nu(\theta)}{\nu_0} = \frac{\nu(\pi/2 + \theta) - \nu(\theta)}{\nu_0} \sim \frac{\Delta \bar{c}_\theta}{c} \tag{6.32}$$

where ν_0 is the reference frequency of the two resonators.

Within the above basic scheme, let us see how experimental results are presented. For instance let us consider the recent work of Nagel et al. [24] where an *average* fractional anisotropy $|\langle \frac{\Delta \bar{c}_\theta}{c} \rangle| \lesssim 10^{-18}$ was reported. With this new result, by looking at their Figure 1 where all ether-drift experiments are combined, one gets the impression of a steady, substantial improvement over the original 1887 Michelson-Morley result $\frac{|\Delta \bar{c}_\theta|}{c} \lesssim 10^{-9}$. All together, their plot supports the traditional view of a series of null results with better and better systematics.

However, our re-analysis in Chap.3 and Chap.5 has confirmed the claims of those experts as Hicks and Miller [6,7] that have seriously questioned the traditional null interpretation of the very early measurements. In their opinion, the small residuals should not be neglected. Therefore one may wonder if, indeed, this first impression is correct. Indeed, the various measurements were performed in different conditions, i.e. with light propagating in gaseous media (as in [1,7,21,22,102]) or in a high vacuum (as in [25–29]) or inside dielectrics with a large refractive index (as in [24,30]) and there could be physical reasons which prevent a straightforward comparison. In this case, the difference between old experiments (in air or gaseous helium) and modern experiments (in vacuum or solid dielectrics) might not depend on the technological progress only but also on the different media that were tested.

In particular, for ether-drift experiments where light propagates in air at atmospheric pressure and at room temperature our analysis predicts a light anisotropy $\frac{\Delta \bar{c}_\theta}{c} = \mathcal{O}(10^{-10})$ which is precisely the value experimentally observed. Then, why should such experiments be considered in the same plot with experiments performed in the cryogenic regime and/or in a very high vacuum?

Here, to clarify the various aspects, what we need is a modern version of Maxwell's calculation which replaces his original estimate. This will be presented in the following section for the case of dielectric media whose refractive index is infinitesimally close to unity (experiments in vacuum and/or in solid dielectrics are postponed to the final Chap.7).

6.4   A modern version of Maxwell's calculation

By leaving out the first few ζ's as free parameters in the fits, Eq.(6.30) can already represent a viable form to compare with experiments. Still, one can further sharpen the predictions by exploiting one more derivation of the $\epsilon \to 0$ limit with a preferred frame. This other argument is based on the effective space-time metric $g^{\mu\nu} = g^{\mu\nu}(\mathcal{N})$ which, through the relation $g^{\mu\nu} p_\mu p_\nu = 0$, describes light propagation in a medium of refractive index $\mathcal{N}$, see e.g. [160] and references quoted therein. For the quantum theory, a derivation of this metric from first principles was given by Jauch and Watson [161] who worked out the quantization of the electromagnetic field in a dielectric. They noticed that the procedure introduces unavoidably a preferred reference frame, the one where the photon energy spectrum does not depend on the direction of propagation, and which is "usually taken as the system for which the medium is at rest". However, such an identification reflects the point of view of special relativity with no preferred frame. Instead, one can adapt their results to the case where the angle-independence of the photon energy is only valid when both medium and observer are at rest in some particular frame Σ.

With this premise, let us consider two identical optical cavities, namely cavity 1, at rest in Σ, and cavity 2, at rest in S', and denote by $\pi_\mu \equiv (\frac{E_\pi}{c}, \pi)$ the light 4-momentum for Σ in his cavity 1 and by $p_\mu \equiv (\frac{E_p}{c}, \mathbf{p})$ the corresponding light 4-momentum for S' in his cavity 2. Let us also denote by $g^{\mu\nu}$ the space-time metric that S' uses in the relation $g^{\mu\nu}p_\mu p_\nu = 0$ and by

$$\gamma^{\mu\nu} = \mathrm{diag}(\mathcal{N}^2, -1, -1, -1) \tag{6.33}$$

the metric used by Σ in the relation $\gamma^{\mu\nu}\pi_\mu\pi_\nu = 0$ and which gives an isotropic velocity $c_\gamma = E_\pi/|\pi| = \frac{c}{\mathcal{N}}$. Notice that, in this framework, special relativity is included as a particular case where there is no observable difference between Σ and S' and the two frames are placed on the same footing.

Let us first consider the ideal vacuum limit $\mathcal{N} = 1$. Here, the frame independence of the velocity of light requires to impose

$$g^{\mu\nu}(\mathcal{N} = 1) = \gamma^{\mu\nu}(\mathcal{N} = 1) = \eta^{\mu\nu} \tag{6.34}$$

where $\eta^{\mu\nu}$ is the Minkowski tensor. This standard equality amounts to introduce a transformation matrix, say A^μ_ν, such that

$$g^{\mu\nu} = A^\mu_\rho A^\nu_\sigma \eta^{\rho\sigma} = \eta^{\mu\nu}. \tag{6.35}$$

This relation is strictly valid for $\mathcal{N} = 1$. However, by continuity, one is driven to conclude that an analogous relation between $g^{\mu\nu}$ and $\gamma^{\mu\nu}$ should also hold in the $\epsilon \to 0$ limit. The only subtlety is that relation (6.35) does not fix uniquely A^μ_ν. In fact, one can either choose the identity matrix, i.e. $A^\mu_\nu = \delta^\mu_\nu$, or a Lorentz transformation, i.e. $A^\mu_\nu = \Lambda^\mu_\nu$. Since for any finite v these two matrices cannot be related by an infinitesimal transformation, it follows that A^μ_ν is a two-valued function in the $\epsilon \to 0$ limit.

Therefore, in principle, there are two solutions. Namely, if A^μ_ν is the identity matrix, we expect a first solution

$$[g^{\mu\nu}(\mathcal{N})]_1 = \gamma^{\mu\nu} \sim \eta^{\mu\nu} + 2\epsilon\delta^\mu_0\delta^\nu_0 \tag{6.36}$$

while, if A^μ_ν is a Lorentz transformation, we expect the other solution

$$[g^{\mu\nu}(\mathcal{N})]_2 = \Lambda^\mu_\rho\Lambda^\nu_\sigma\gamma^{\rho\sigma} \sim \eta^{\mu\nu} + 2\epsilon v^\mu v^\nu \tag{6.37}$$

v_μ being the dimensionless S' 4-velocity, $v_\mu \equiv (v_0, \mathbf{v}/c)$ with $v_\mu v^\mu = 1$.

Notice that with the former choice, implicitly adopted in special relativity to preserve isotropy in all reference systems also for $\mathcal{N} \neq 1$, one is introducing a discontinuity in the transformation matrix for any $\epsilon \neq 0$. Indeed, the whole emphasis on Lorentz transformations depends on enforcing Eq.(6.35) for $A^\mu_\nu = \Lambda^\mu_\nu$ so that $\Lambda^{\mu\sigma}\Lambda^\nu_\sigma = \eta^{\mu\nu}$ and the Minkowski metric applies to all equivalent frames.

On the other hand, with the latter solution, by replacing in the relation $p_\mu p_\nu g^{\mu\nu} = 0$, the photon energy now depends on the direction of propagation. By introducing $\kappa = \mathcal{N}^2 - 1 \sim 2\epsilon$ one finds

$$E(|\mathbf{p}|, \theta) = c \, \frac{-\kappa v_0 \sigma + \sqrt{|\mathbf{p}|^2(1 + \kappa v_0^2) - \kappa \sigma^2}}{1 + \kappa v_0^2} \tag{6.38}$$

with $v_0^2 = 1 + \mathbf{v}^2/c^2$ and

$$\sigma = \mathbf{p} \cdot \frac{\mathbf{v}}{c} = |\mathbf{p}|\beta \cos\theta. \tag{6.39}$$

Here $\beta = \frac{|\mathbf{v}|}{c}$ and $\theta \equiv \theta_{\text{lab}}$ indicates the angle defined, in the laboratory S' frame, between the photon momentum and $\mathbf{v}$. By using the above relation, one gets the one-way velocity of light

$$\frac{E(|\mathbf{p}|, \theta)}{|\mathbf{p}|} = c_\gamma(\theta) = c \, \frac{-\kappa\beta\sqrt{1 + \beta^2}\cos\theta + \sqrt{1 + \kappa + \kappa\beta^2 \sin^2\theta}}{1 + \kappa(1 + \beta^2)}. \tag{6.40}$$

or to $\mathcal{O}(\epsilon)$ and $\mathcal{O}(\beta^2)$

$$c_\gamma(\theta) \sim \frac{c}{\mathcal{N}} \left[1 - 2\epsilon\beta\cos\theta - \epsilon\beta^2(1 + \cos^2\theta)\right]. \tag{6.41}$$

From this one can compute the two-way velocity[4]

$$\bar{c}_\gamma(\theta) = \frac{2c_\gamma(\theta)c_\gamma(\pi + \theta)}{c_\gamma(\theta) + c_\gamma(\pi + \theta)}$$

$$\sim \frac{c}{\mathcal{N}} \left[1 - \epsilon\beta^2\left(1 + \cos^2\theta\right)\right] \tag{6.42}$$

---

[4]There is a subtle difference between our Eqs.(6.41) and(6.42) and the corresponding Eqs. (6) and (10) of ref. [104] that has to do with the relativistic aberration of the angles. Namely, in ref. [104], with the (wrong) motivation that the anisotropy is $\mathcal{O}(\beta^2)$, no attention was paid to the precise definition of the angle between the Earth's velocity and the direction of the photon momentum. Thus the two-way speed of light in the S' frame was parameterized in terms of the angle $\theta \equiv \theta_\Sigma$ as seen in the Σ frame. This can be explicitly checked by replacing in our Eqs. (6.41) and(6.42) the aberration relation $\cos\theta_{\text{lab}} = (-\beta + \cos\theta_\Sigma)/(1 - \beta\cos\theta_\Sigma)$ or equivalently by replacing $\cos\theta_\Sigma = (\beta + \cos\theta_{\text{lab}})/(1 + \beta\cos\theta_{\text{lab}})$ in Eqs. (6) and (10) of Ref. [104]. However, the apparatus is at rest in the laboratory frame, so that the correct orthogonality condition of two optical cavities at angles θ and $\pi/2 + \theta$ is expressed in terms of $\theta = \theta_{\text{lab}}$ and not in terms of $\theta = \theta_\Sigma$. This trivial remark produces however a non-trivial difference. In fact, the final anisotropy is now smaller by a factor of 3 than the one computed in Ref. [104] by adopting the wrong definition of orthogonality in terms of $\theta = \theta_\Sigma$.

Eq.(6.42) corresponds to setting in Eq.(6.30) $\zeta_0 = 4/3$, $\zeta_2 = 2/3$ and all $\zeta_{2n} = 0$ for $n > 1$ and can be considered a modern version of Maxwell's original calculation. It represents a definite, alternative model for the interpretation of experiments performed close to the ideal vacuum limit $\epsilon = 0$, such as in gaseous systems, and will be adopted in the following.

A conceptual detail concerns the relation of the gas refractive index $\mathcal{N}$, as introduced in Eq.(6.33), to the experimental quantity $\mathcal{N}_{\mathrm{exp}}$ which is extracted from measurements of the two-way velocity in the earth laboratory. In terms of the θ-dependent refractive index Eq.(6.17) one should thus define the experimental value by an angular average of Eq.(6.42), i.e.

$$\frac{c}{\mathcal{N}_{\mathrm{exp}}} \equiv \langle \frac{c}{\mathcal{N}(\theta)} \rangle_\theta = \frac{c}{\mathcal{N}} \left[1 - \frac{3}{2}(\mathcal{N} - 1)\beta^2 \right]. \tag{6.43}$$

From this relation, one can determine in principle the unknown value $\mathcal{N} \equiv \mathcal{N}(\Sigma)$ (as if the container of the gas were at rest in Σ), in terms of the experimentally known quantity $\mathcal{N}_{\mathrm{exp}} \equiv \mathcal{N}(earth)$ and of v. For instance, for air the most precise determinations are at the level 10^{-7}, say $\mathcal{N}_{\mathrm{exp}} = 1.0002924..$ for light of 589 nm, at 0 °C and atmospheric pressure. In practice, for the standard velocity values involved in most cosmic motions, say $v \sim 300$ km/s, the difference between $\mathcal{N}(\Sigma)$ and $\mathcal{N}(earth)$ is below 10^{-9} and thus completely negligible. The same holds true for the other gaseous systems (say nitrogen, carbon dioxide, helium,..) for which the present experimental accuracy in the refractive index is, at best, at the level 10^{-7}. Finally, the isotropic two-way speed of light is better determined in the low-pressure limit where $(\mathcal{N} - 1) \to 0$. In the same limit, for any given value of v, the approximation $\mathcal{N}(\Sigma) = \mathcal{N}(earth)$ becomes better and better.

From Eq.(6.42) we obtain a fractional anisotropy

$$\frac{\Delta \bar{c}_\theta}{c} = \frac{\bar{c}_\gamma(\pi/2 + \theta) - \bar{c}_\gamma(\theta)}{c} \sim \epsilon \frac{v^2}{c^2} \cos 2(\theta - \theta_0). \tag{6.44}$$

Here v and θ_0 are respectively the magnitude and the direction of the drift in the interferometer's plane so that, from Eq.(6.29), one finds directly the fringe pattern

$$\frac{\Delta \lambda(\theta)}{\lambda} = \frac{2D}{\lambda} \frac{\Delta \bar{c}_\theta}{c} \sim 2\epsilon \frac{D}{\lambda} \frac{v^2}{c^2} \cos 2(\theta - \theta_0). \tag{6.45}$$

In this scheme, as anticipated, the ether drift is a pure 2nd-harmonic effect, i.e. periodic in the range $[0, \pi]$, as in the classical theory (see e.g. [3]) . Only its amplitude

$$A_2 = 2\epsilon \frac{D}{\lambda} \frac{v^2}{c^2} \tag{6.46}$$

is suppressed by the very small factor 2ϵ with respect to the classical prediction $A_2^{\text{class}} = \frac{D}{\lambda}\frac{v^2}{c^2}$. Thus one can re-absorb all effects by expressing

$$A_2 = \frac{D}{\lambda}\frac{v_{\text{obs}}^2}{c^2} \qquad (6.47)$$

in terms of an *observable* velocity

$$v_{\text{obs}}^2 \sim 2\epsilon v^2 \qquad (6.48)$$

which depends on the gas refractive index and is the one traditionally reported in the classical analysis of the data.

However, as anticipated in the Introduction, for a proper comparison with experiments a change of perspective is needed in the physical description of the ether-drift phenomenon. In the following section, we will illustrate a simple stochastic model that we propose for the analysis of the data.

6.5 A stochastic form of ether-drift

To make explicit the time dependence of the signal let us re-write Eq.(6.44) as

$$\frac{\Delta \bar{c}_\theta(t)}{c} \sim \epsilon \frac{v^2(t)}{c^2} \cos 2(\theta - \theta_0(t)) \qquad (6.49)$$

where $v(t)$ and $\theta_0(t)$ indicate respectively the instantaneous magnitude and direction of the drift in the plane of the interferometer. This can also be re-written as

$$\frac{\Delta \bar{c}_\theta(t)}{c} \sim 2S(t)\sin 2\theta + 2C(t)\cos 2\theta \qquad (6.50)$$

with

$$2C(t) = \epsilon \frac{v_x^2(t) - v_y^2(t)}{c^2} \qquad 2S(t) = \epsilon \frac{2v_x(t)v_y(t)}{c^2} \qquad (6.51)$$

and $v_x(t) = v(t)\cos\theta_0(t)$, $v_y(t) = v(t)\sin\theta_0(t)$.

As anticipated in the Introduction, the standard assumption to analyze the data is based on the idea of smooth, regular modulations of the signal associated with a cosmic earth velocity. In general, this is characterized by a magnitude V, a right ascension α and an angular declination γ. These parameters can be considered constant for short-time observations of a few days where there are no appreciable changes due to the earth orbital velocity around the sun. In this framework, where the only time dependence is due to the earth rotation, the traditional identifications are $v(t) \equiv \tilde{v}(t)$ and

$\theta_0(t) \equiv \tilde{\theta}_0(t)$ where $\tilde{v}(t)$ and $\tilde{\theta}_0(t)$ derive from the simple application of spherical trigonometry [133]

$$\cos z(t) = \sin\gamma\sin\phi + \cos\gamma\cos\phi\cos(\tau - \alpha) \qquad (6.52)$$

$$\tilde{v}(t) = V\sin z(t) \qquad (6.53)$$

$$\tilde{v}_x(t) = \tilde{v}(t)\cos\tilde{\theta}_0(t) = V\left[\sin\gamma\cos\phi - \cos\gamma\sin\phi\cos(\tau - \alpha)\right] \qquad (6.54)$$

$$\tilde{v}_y(t) = \tilde{v}(t)\sin\tilde{\theta}_0(t) = V\cos\gamma\sin(\tau - \alpha). \qquad (6.55)$$

Here $z = z(t)$ is the zenithal distance of $\mathbf{V}$, ϕ is the latitude of the laboratory, $\tau = \omega_{\text{sid}}t$ is the sidereal time of the observation in degrees $(\omega_{\text{sid}} \sim \frac{2\pi}{23^h 56'})$ and the angle θ_0 is counted conventionally from North through East so that North is $\theta_0 = 0$ and East is $\theta_0 = 90°$. With the identifications $v(t) \equiv \tilde{v}(t)$ and $\theta_0(t) \equiv \tilde{\theta}_0(t)$, one then arrives to the identifications $S(t) \equiv \tilde{S}(t)$ and $C(t) \equiv \tilde{C}(t)$ where $\tilde{S}(t)$ and $\tilde{C}(t)$ have the simple Fourier decomposition

$$\tilde{S}(t) = S_{s1}\sin\tau + S_{c1}\cos\tau + S_{s2}\sin(2\tau) + S_{c2}\cos(2\tau) \qquad (6.56)$$

$$\tilde{C}(t) = C_0 + C_{s1}\sin\tau + C_{c1}\cos\tau + C_{s2}\sin(2\tau) + C_{c2}\cos(2\tau). \qquad (6.57)$$

Here the C_k and S_k Fourier coefficients depend on the three parameters (V, α, γ). By defining $\mathcal{R} \equiv \frac{D_\epsilon}{\lambda}\frac{V^2}{c^2}$, one finds

$$C_0 = -\frac{1}{4}\mathcal{R}(3\cos 2\gamma - 1)\cos^2\phi \qquad (6.58)$$

$$C_{s1} = -\frac{1}{2}\mathcal{R}\sin\alpha\sin 2\gamma\sin 2\phi \;\; ; \;\; C_{c1} = -\frac{1}{2}\mathcal{R}\cos\alpha\sin 2\gamma\sin 2\phi \qquad (6.59)$$

$$C_{s2} = \frac{1}{2}\mathcal{R}\sin 2\alpha\cos^2\gamma(1 + \sin^2\phi) \;\; ; \;\; C_{c2} = \frac{1}{2}\mathcal{R}\cos 2\alpha\cos^2\gamma(1 + \sin^2\phi) \qquad (6.60)$$

$$S_{s1} = -\frac{C_{c1}}{\sin\phi} \;\; ; \;\; S_{c1} = \frac{C_{s1}}{\sin\phi} \qquad (6.61)$$

$$S_{s2} = -\frac{2\sin\phi}{1 + \sin^2\phi}C_{c2} \;\; ; \;\; S_{c2} = \frac{2\sin\phi}{1 + \sin^2\phi}C_{s2}. \qquad (6.62)$$

Though, the identification of the instantaneous quantities $v_x(t)$ and $v_y(t)$ with their counterparts $\tilde{v}_x(t)$ and $\tilde{v}_y(t)$ is not necessarily true. In fact, by comparing with the motion of a body in a fluid, this identification amounts to assume a form of regular, laminar flow where the microscopic velocity field, which affects light propagation in the laboratory, coincides with the macroscopic velocity field as determined by the earth cosmic motion. Instead, as anticipated in Chap.1, by starting from the intuitive idea of the physical vacuum as a fluid with zero-viscosity (or with infinite Reynolds number for any velocity of the flow), one could consider the alternative situation where the velocity field is a non-differentiable function [42,43]. Thus, if the ordinary formulation in terms of differential equations breaks down, one has to adopt some other description, for instance a formulation in terms of random Fourier series [42,162,163]. In this other approach, the parameters of the macroscopic motion are used to fix the typical boundaries for a microscopic velocity field which has an intrinsic non-deterministic nature. Altogether, this leads to the idea of the physical vacuum as a fundamental stochastic medium in agreement with some basic foundational aspects [45] of both quantum physics and relativity[5].

The simplest model, adopted in refs. [38,45], corresponds to a turbulence which, at small scales, appears homogeneous and isotropic. The analysis of the previous section, can then be embodied in an effective space-time metric for light propagation

$$g^{\mu\nu}(t) \sim \eta^{\mu\nu} + 2\epsilon v^{\mu}(t)v^{\nu}(t) \qquad (6.63)$$

where $v^{\mu}(t)$ is a random 4-velocity field which describes the drift and whose boundaries depend on a smooth field $\tilde{v}^{\mu}(t)$ determined by the average earth motion. If this corresponds to the actual physical situation, it is easy to see why a genuine stochastic signal can become consistent with average values $(C_k)^{\text{avg}} = (S_k)^{\text{avg}} = 0$ obtained by fitting the data with Eqs.(6.56) and (6.57).

---

[5]At the end of Chapt.2 we have recalled the old picture of the ether as a turbulent fluid. In this derivation, Lorentz covariance of Maxwell equations is not postulated from scratch but is a symmetry emerging from an underlying physical system whose constituents obey classical mechanics. More recently, the turbulent-ether model has been re-formulated by Troshkin [90] (see also [91] and [92]) in the framework of the Navier-Stokes equation and by Saul [164] by starting from Boltzmann's transport equation. As another example, the same picture of the physical vacuum (or ether) as a turbulent fluid was Nelson's [165] starting point. In particular, the zero-viscosity limit gave him the motivation to expect that "the Brownian motion in the ether will not be smooth" and, therefore, to conceive the particular form of kinematics which is at the base of his stochastic derivation of the Schrödinger equation. A qualitatively similar picture is also obtained by representing relativistic particle propagation from the superposition, at very short time scales, of non-relativistic particle paths with different Newtonian mass [166]. In this formulation, particles randomly propagate (in the sense of Brownian motion) in an underlying granular medium which replaces the trivial empty vacuum [167]. For more details, see [45].

Our intention is to simulate the two components of the velocity in the
x-y plane, at a given fixed location in the laboratory, to reproduce the $S(t)$
and $C(t)$ functions Eq.(6.51). For a homogeneous turbulence, one finds the
general expressions

$$v_x(t) = \sum_{n=1}^{\infty} [x_n(1) \cos \omega_n t + x_n(2) \sin \omega_n t] \qquad (6.64)$$

$$v_y(t) = \sum_{n=1}^{\infty} [y_n(1) \cos \omega_n t + y_n(2) \sin \omega_n t] \qquad (6.65)$$

where $\omega_n = 2n\pi/T$, T being a time scale which represents a common
period of all stochastic components. For numerical simulations, the typical
value $T = T_{\text{day}} = 24$ hours was adopted [38, 45]. However, it was also
checked with a few runs that the statistical distributions of the various
quantities do not change substantially by varying T in the rather wide
range $0.1\, T_{\text{day}} \leq T \leq 10\, T_{\text{day}}$.

The coefficients $x_n(i = 1, 2)$ and $y_n(i = 1, 2)$ are random variables with
zero mean and have the physical dimension of a velocity. In general, we can
denote by $[-d_x(t), d_x(t)]$ the range for $x_n(i = 1, 2)$ and by $[-d_y(t), d_y(t)]$
the corresponding range for $y_n(i = 1, 2)$. Statistical isotropy would require
to impose $d_x(t) = d_y(t)$. However, to illustrate the more general case, let
us first consider $d_x(t) \neq d_y(t)$. In terms of these boundaries, the only
non-vanishing (quadratic) statistical averages are

$$\langle x_n^2(i = 1, 2) \rangle_{\text{stat}} = \frac{d_x^2(t)}{3\, n^{2\eta}} \qquad \langle y_n^2(i = 1, 2) \rangle_{\text{stat}} = \frac{d_y^2(t)}{3\, n^{2\eta}} \qquad (6.66)$$

in a uniform-probability model within the intervals $[-d_x(t), d_x(t)]$ and
$[-d_y(t), d_y(t)]$. Here, the exponent η controls the power spectrum of the
fluctuating components. For numerical simulations, between the two values
$\eta = 5/6$ and $\eta = 1$ reported in ref. [163], we have adopted $\eta = 1$ which
corresponds to the Lagrangian description where the point of measurement
is a wandering material point in the fluid.

Finally, the connection with the earth cosmic motion is obtained by
identifying $d_x(t) = \tilde{v}_x(t)$ and $d_y(t) = \tilde{v}_y(t)$ as given in Eqs. (6.52)−(6.55).
In this case, it is natural to adopt the set $V = 370$ km/s, $\alpha = 168$ degrees,
$\gamma = -7$ degrees, which describes the average earth motion with respect to
the CMB.

If, however, we require statistical isotropy, the relation

$$\tilde{v}_x^2(t) + \tilde{v}_y^2(t) = \tilde{v}^2(t) \qquad (6.67)$$

requires the identification[6]

$$d_x(t) = d_y(t) = \frac{\tilde{v}(t)}{\sqrt{2}}. \tag{6.68}$$

For such isotropic model, by combining Eqs.(6.64)−(6.68) and in the limit of an infinite statistics, one gets

$$\langle v_x^2(t) \rangle_{\text{stat}} = \langle v_y^2(t) \rangle_{\text{stat}} = \frac{\tilde{v}^2(t)}{2} \frac{1}{3} \sum_{n=1}^{\infty} \frac{1}{n^2} = \frac{\tilde{v}^2(t)}{2} \frac{\pi^2}{18}$$

$$\langle v_x(t) v_y(t) \rangle_{\text{stat}} = 0 \tag{6.69}$$

and vanishing statistical averages

$$\langle C(t) \rangle_{\text{stat}} = 0 \qquad\qquad \langle S(t) \rangle_{\text{stat}} = 0 \tag{6.70}$$

at *any* time t, see Eqs.(6.51). Therefore, by construction, this model gives a definite non-zero signal but, if the same signal were fitted with Eqs.(6.56) and (6.57), it also gives average values $(C_k)^{\text{avg}} = 0$, $(S_k)^{\text{avg}} = 0$ for the Fourier coefficients.

6.6 Reconsidering the classical experiments

To fully appreciate the change of perspective, let us consider the traditional procedure of data taking in the classical experiments. Fringe shifts were observed at the same sidereal time on a few consecutive days so that changes in the earth orbital velocity could be ignored. Then, see Eqs.(6.45) and (6.50), the data were averaged at any given angle θ

$$\langle \frac{\Delta\lambda(\theta;t)}{\lambda} \rangle_{\text{stat}} = \frac{2D}{\lambda} [2 \sin 2\theta \, \langle S(t) \rangle_{\text{stat}} + 2 \cos 2\theta \, \langle C(t) \rangle_{\text{stat}}] \tag{6.71}$$

and these averages were compared with various models of cosmic motion.

But, if the ether-drift is a genuine stochastic phenomenon, as expected if the physical vacuum were similar to a turbulent fluid which becomes isotropic at small scales, these average combinations should vanish *exactly* for an infinite number of measurements. Thus, averages of vectorial quantities are non vanishing just because the statistics is finite and forming the averages Eq.(6.71) is not a meaningful procedure. In particular, the direction $\theta_0(t)$ of the drift in the plane of the interferometer (defined by the

---

[6]The correct normalization Eq.(6.67) produces boundaries which are smaller by a factor $\frac{1}{\sqrt{2}}$ as compared to those of ref. [45] where the relation $d_x(t) = d_y(t) \sim \tilde{v}(t)$ was assumed. For this reason, in view of Eqs.(6.66), the resulting amplitudes of the signal are now predicted to be smaller by about a factor of 2.

relation $\tan 2\theta_0(t) = S(t)/C(t))$ is a completely random quantity which has no definite limit by combining a large number of observations. Instead, one should concentrate on the 2nd-harmonic amplitudes

$$A_2(t) = \frac{2D}{\lambda} \, 2\sqrt{S^2(t) + C^2(t)}. \tag{6.72}$$

These are positive-definite quantities and, as such, remain definitely non-zero after any averaging procedure. In addition, being rotationally invariant, their statistical properties remain the same by adopting the isotropic model Eq.(6.68) or the non-isotropic choice $d_x(t) \equiv \tilde{v}_x(t)$ and $d_y(t) \equiv \tilde{v}_y(t)$.

As a matter of fact, by restricting to the amplitudes, one finds a good consistency of the data of the classical experiments with the kinematical parameters obtained from the observations of the CMB.

For instance, let us consider the average 2nd-harmonic amplitude $A_2^{\mathrm{EXP}} \sim 0.016 \pm 0.006$ extracted from the six sessions of the Michelson-Morley experiment (see Chap.3). By comparing with the classical prediction $A_2^{\mathrm{class}} = \frac{D}{\lambda} \frac{(30\mathrm{km/s})^2}{c^2} \sim 0.20$, this average amplitude corresponds to an observable velocity $v_{\mathrm{obs}} \sim (8.4 \pm 1.7)$ km/s. Therefore, by using Eq.(6.48), for air at room temperature and atmospheric pressure where $\epsilon \sim 2.8 \cdot 10^{-4}$, one obtains a true kinematical value

$$v \sim (355 \pm 70) \text{ km/s} \qquad (\text{Michelson} - \text{Morley}) \qquad (6.73)$$

Notice the consistency with the determination $v \sim 370$ km/s obtained from the observations of the CMB.

Similar considerations can be done for the Morley-Miller 1902−1905 observations. As explained in Chapt.5, after that Miller corrected the wrong combination of data performed in the original articles, the resulting range $v_{\mathrm{obs}} \sim (8.5 \pm 1.5)$ km/s, with Eq.(6.48), becomes

$$v \sim (359 \pm 62) \text{ km/s} \qquad (\text{Morley} - \text{Miller}) \qquad (6.74)$$

Analogously, let us consider Miller's extensive observations. As discussed in Chap.5, after the critical re-analysis of his original measurements performed by the Shankland team [47], one has an accurate determination of the overall average for all epochs of the year (see Table III of [47]) . The resulting amplitude $A_2^{\mathrm{EXP}} = 0.044 \pm 0.022$, when compared with the equivalent classical prediction for Miller's interferometer $A_2^{\mathrm{class}} = \frac{D}{\lambda} \frac{(30\mathrm{km/s})^2}{c^2} \sim 0.56$, gives $v_{\mathrm{obs}} \sim (8.4 \pm 2.2)$ km/s and, by using Eq.(6.48), a true kinematical value

$$v \sim (355 \pm 90) \text{ km/s} \qquad (\text{Miller}) \qquad (6.75)$$

again consistent with the CMB determination.

As anticipated, the perfect identity of these determinations obtained in completely different experimental conditions (in the basement of Cleveland University or on top of Mount Wilson) indicates that an interpretation [47] of the residuals in terms of temperature effects is only acceptable provided they have a *non-local* origin, e.g. the CMB temperature dipole.

Also, the range of observable velocity obtained in Chap.5 from Tomaschek's starlight experiment, namely $v_{obs} = 7.7^{+2.1}_{-2.8}$ km/s, becomes a kinematical velocity

$$v = 325^{+87}_{-116} \text{ km/s} \qquad \text{(Tomaschek)} \qquad (6.76)$$

Analogous considerations can be applied to Kennedy's and Illingworth's experiments in gaseous helium. By ignoring the directional character of the data and replacing $\epsilon \sim 3.3 \cdot 10^{-5}$ in Eq.(6.48), Kennedy's upper $v_{obs} <$ 5 km/s reported in Chap.5 becomes

$$v < 600 \text{ km/s} \qquad \text{(Kennedy)} \qquad (6.77)$$

while, our estimate of Illingworth's observable velocity $v_{obs} \sim 2.4^{+0.8}_{-1.2}$ km/s translates into

$$v \sim 295^{+98}_{-148} \text{ km/s} \qquad \text{(Illingworth)} \qquad (6.78)$$

6.7 Reanalysis of the Piccard-Stahel experiment

In the case of the Piccard-Stahel experiment, the range for the *observable* velocity found in Chapt.5, namely $4 \text{ km/s} \lesssim v_{obs} \lesssim 8 \text{ km/s}$ at the 75 % C.L., would imply, with Eq.(6.48), the corresponding range for the kinematical velocity

$$169 \text{ km/s} \lesssim v \lesssim 338 \text{ km/s} \qquad \text{(Piccard} - \text{Stahel 75\% C.L.)} \quad (6.79)$$

In view of the precision of the Piccard-Stahel experiment we have, however, tried to obtain a more refined check by comparing directly their data with our stochastic model which takes into account random fluctuations of the local velocity components $v_x(t)$ and $v_y(t)$. To this end, it is convenient to re-express the amplitude in the form

$$A_2(t) = \frac{2\epsilon D}{\lambda} \frac{v_x^2(t) + v_y^2(t)}{c^2} \sim 3.6 \cdot 10^{-3} \frac{v_x^2(t) + v_y^2(t)}{(300 \text{ km/s})^2} \qquad (6.80)$$

where we have replaced the numerical values $D/\lambda \sim 6.4 \cdot 10^6$ and $\epsilon \sim 2.8 \cdot 10^{-4}$.

For our analysis, we have first considered the sample of data collected at Mt. Rigi in Switzerland [137] (Long.= 8.50 degrees East, Lat.= 47 degrees North). As reported in our Chap.5, the data taking consisted of 12 sets of measurements, each corresponding to 10 rotations of the interferometer and the 12 amplitudes (in units 10^{-3}) were $A_2^{\text{EXP}} = 3.4, 1.1, 4.0, 2.4, 2.4, 4.3, 2.3, 2.6, 0.6, 2.0, 1.2, 3.9$ with an average amplitude

$$\langle A_2^{\text{EXP}} \rangle = (2.5 \pm 1.2) \cdot 10^{-3} \qquad \text{(average expt. Mt.Rigi amplitude)}.$$
$$(6.81)$$

We have then run our numerical simulation by using the CMB parameters (V, α, γ) in the scalar velocity Eq.(6.53) which determines the boundaries Eq.(6.68) of the stochastic components in Eqs.(6.64) and (6.65).

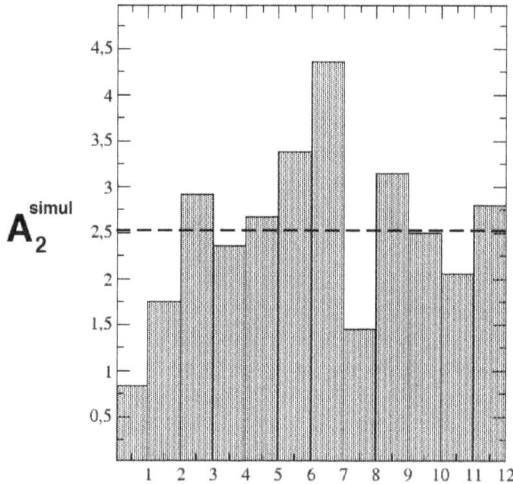

Fig. 6.3   *A single simulation of 12 amplitudes, in units 10^{-3}. Each value corresponds to the average of 10 individual rotations and the time is the same as for the Piccard-Stahel Mt. Rigi measurements. The dashed line indicates the global average $\langle A_2^{\text{simul}} \rangle = 2.51 \cdot 10^{-3}$ of the 12 values. The limits on the random Fourier components in Eqs.(6.64) and (6.65) were fixed by the CMB kinematical parameters as explained in the text.*

To start with, we have generated 10 individual amplitudes (one for each rotation) and computed their average value. Each average then represents one of 12 possible measurements. Among the many simulations, we first tried to select sets of 12 measurements whose global average amplitude is close to the experimental value $\langle A_2^{\text{EXP}} \rangle \sim 2.5 \cdot 10^{-3}$. This first type of search has no systematic nature but is useful to get a qualitative impression of the agreement with our model. A particular set of such 12 measurements, obtained from a single random sequence, and for N=1000 Fourier modes in Eqs.(6.64) and (6.65), is shown in Fig. 6.3 and corresponds to a simulated average amplitude $\langle A_2^{\text{simul}} \rangle = (2.51 \pm 0.93) \cdot 10^{-3}$.

Then, for a more complete analysis, we started to change the seed s of the random sequence, the number N of the Fourier modes and increase the statistics up to factor of ten (i.e. up to a number $n = 120$ of simulated measurements, each deriving again from averaging over 10 individual rotations). The overall average simulated amplitude thus becomes

$$\langle A_2^{\text{simul}} \rangle = (3.0 \pm 1.0) \cdot 10^{-3} \qquad \text{(average simulated Mt.Rigi amplitude)}$$

$$(6.82)$$

where the central value and the error take into account both the effect of varying the triple (s, N, n), as well as the typical standard deviation $0.8 \cdot 10^{-3}$ for a fixed choice (s, N, n) of the parameters of the simulation. One of the higher statistics simulations for $n = 120$ is shown in Fig. 6.4.

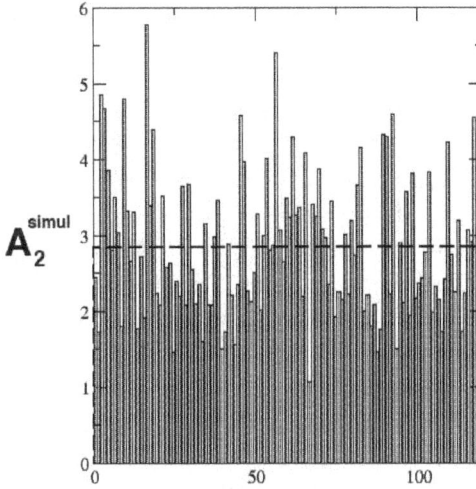

Fig. 6.4   *A single simulation of 120 amplitudes, in units* 10^{-3}. *Each value corresponds to the average of 10 individual rotations and the time is the same as for the Piccard-Stahel Mt. Rigi measurements. The dashed line indicates the average* $\langle A_2^{\text{simul}} \rangle = 2.84 \cdot 10^{-3}$ *of the 120 values. The limits on the random Fourier components in Eqs.(6.64) and (6.65) were fixed by the CMB kinematical parameters as explained in the text.*

From this analysis, we can draw the following conclusions. On the one hand, both in the low-statistics and higher-statistics simulations, the characteristic scatter of the data is reproduced in a satisfactory way. On the other hand, by assuming the CMB kinematical parameters, the projection Eq.(6.53) of the cosmic earth velocity at the time of the observations and at the latitude of the laboratory was $\tilde{v}(t) \sim 364$ km/s. Thus, the agreement between the results of our simulations and the Piccard-Stahel experimental amplitudes is actually *much better* than what could be expected from Eq.(6.79). The point is that Eq.(6.79) simply translates the observable velocity into a kinematical velocity by means of the factor 2ϵ in Eq.(6.48) as if the velocity field were a smooth function. Instead, we now find that the agreement improves when one takes into account the stochastic fluctuations.

The simple explanation for such improvement is obtained when one takes a full statistical average and use Eq.(6.69) in Eq.(6.80). This gives

$$\langle A_2(t) \rangle_{\text{stat}} \sim 3.6 \cdot 10^{-3} \; \frac{\langle v_x^2(t) + v_y^2(t) \rangle_{\text{stat}}}{(300 \text{ km/s})^2} \sim 3.6 \cdot 10^{-3} \; \frac{\pi^2}{18} \; \frac{\tilde{v}^2(t)}{(300 \text{ km/s})^2}$$
(6.83)

In this way the prediction for the average amplitude, in the limit of an infinite statistics, becomes

$$\langle A_2(t) \rangle_{\text{stat}} \sim 2.0 \cdot 10^{-3} \; \frac{\tilde{v}^2(t)}{(300 \text{ km/s})^2}$$
(6.84)

so that, from the same data, one gets values of the scalar projection $\tilde{v}(t)$ which are larger by a factor $\sqrt{18/\pi^2} \sim 1.35$ as compared to the smooth prediction where one just replaces $(v_x^2(t) + v_y^2(t))$ with $(\tilde{v}_x^2(t) + \tilde{v}_y^2(t)) = \tilde{v}^2(t)$ in Eq. (6.80). Therefore, by using Eq. (6.84) the Mt.Rigi amplitude $\langle A_2^{\text{EXP}} \rangle = (2.5 \pm 1.2) \cdot 10^{-3}$ yields now $\tilde{v}(t) \sim 338^{+72}_{-93}$ km/s, rather than $\tilde{v}(t) \sim 250^{+53}_{-69}$ km/s.

Analogous conclusions can be drawn for the other two sets of 6 measurements which were performed in Brussels, see Chapt.5. We do not repeat the analysis but just limit to quote the average simulated amplitudes $\langle A_2^{\text{simul}} \rangle = (2.8 \pm 0.8) \cdot 10^{-3}$ (midnight) and $\langle A_2^{\text{simul}} \rangle = (2.7 \pm 0.8) \cdot 10^{-3}$ (morning). These simulated amplitudes can now be understood by replacing the average input $\tilde{v}(t) \sim 348$ km/s for the two set of observations in Eq.(6.84) which, for the limit of an infinite statistics, yields $\langle A_2(t) \rangle_{\text{stat}} \sim 2.7 \cdot 10^{-3}$. This prediction is well consistent with the two experimental averages $\langle A_2^{\text{EXP}} \rangle = (4.3 \pm 1.5) \cdot 10^{-3}$ and $\langle A_2^{\text{EXP}} \rangle = (2.1 \pm 1.2) \cdot 10^{-3}$ obtained respectively from the midnight and morning Piccard-Stahel measurements.

Since the two scalar velocities are very close, namely $\tilde{v}(t) \sim 348$ km/s and $\tilde{v}(t) \sim 364$ km/s respectively for Brussels and Mt. Rigi observations, we can neglect the small difference between the two laboratories and replace the average value 356 km/s in Eq.(6.84). By inserting the typical standard deviation of the mean found in our simulations, about $0.8 \cdot 10^{-3}$, we would then conclude that our stochastic model, in the limit of an infinite statistics, predicts an average amplitude for the Piccard-Stahel experiment

$$\langle A_2(t) \rangle_{\text{stat}} \sim (2.8 \pm 0.8) \cdot 10^{-3} \qquad \text{(stochastic model)} \qquad (6.85)$$

This should be compared with the global average obtained in Chapt.5 from the 24 Piccard-Stahel experimental amplitudes

$$\langle A_2^{\text{EXP}} \rangle = (2.8 \pm 1.5) \cdot 10^{-3} \qquad \text{(global Mt.Rigi and Brussels average)} \qquad (6.86)$$

This excellent agreement supports, at the same time, the validity of our stochastic model and the choice of the CMB kinematical parameters to fix the boundaries of the random velocity components in Eqs.(6.64) and (6.65).

6.8 Reanalysis of the MPP experiment

To re-analyze the Michelson-Pease-Pearson (MPP) experiment, we first re-call that, as discussed in Chapt.5, no numerical results are reported in the original articles [139,140]. Instead, for more precise indications, one should look at Pease's article [141]. There one learns that they concentrated on a purely differential type of measurements. This means that they were first averaging the fringe shifts obtained from a large number of observations performed at those sidereal times which, presumably, were corresponding to maxima and minima of the ether-drift effect. Then, these average fringe shifts (at the presumed maxima and minima) were subtracted from each other. These differences, which are the only numbers reported by Pease, were typically about ± 0.004 or smaller. This is also the order of magnitude that is usually compared [47] with the expected classical amplitudes $A_2^{\text{class}} \sim 0.45$ or $A_2^{\text{class}} \sim 0.29$ for optical paths of eighty-five or fifty-five feet respectively.

Now our stochastic model predicts exactly zero averages for vector quantities such as the fringe shifts. Therefore, it would be trivial to reproduce such extremely small values in a numerical simulation of sufficiently high statistics. We have thus decided to compare instead with the only basic experimental session which is reported by Pease (for optical path of fifty-five feet) which, as discussed in Chapt.5, was indicating an experimental

2nd-harmonic amplitude $A_2^{\text{EXP}} \sim 0.006$. By comparing with the classical prediction for 30 km/s, namely $A_2^{\text{class}} \sim 0.29$, this would correspond to an observable velocity $v_{\text{obs}} \sim 4.3$ km/s and, on the basis of Eq.(6.48), to a kinematical velocity of about 180 km/s.

As explained in the previous section, however, by taking into account stochastic fluctuations of the velocity field the kinematical velocity obtained from a given 2nd-harmonic amplitude becomes larger by about a factor $\sqrt{18/\pi^2} \sim 1.35$ with respect to the corresponding prediction with a smooth velocity field. Thus, in our model, this single Pease session may become consistent with a kinematical velocity as large as 245 km/s.

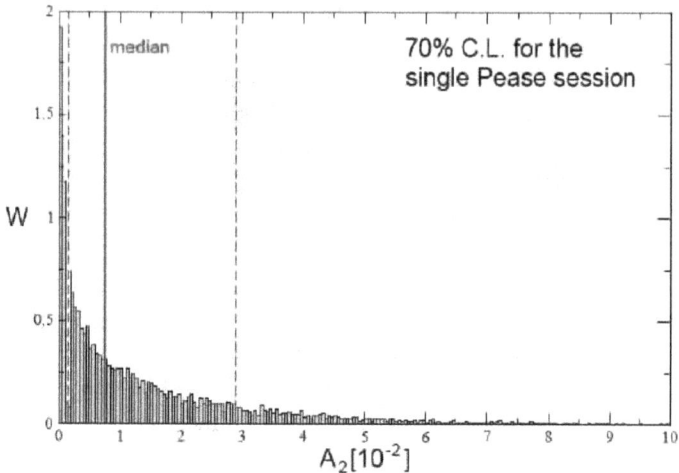

Fig. 6.5   *The histogram W of a numerical simulation of 10,000 instantaneous amplitudes for the single session of January 13, 1928, reported by Pease [141]. The vertical normalization is to a unit area. We show the median and the 70% CL. The limits on the random Fourier components in Eqs.(6.64) and (6.65) were fixed by the CMB kinematical parameters as explained in the text.*

However, to obtain a better understanding of the probability content of that single measurement we have performed a direct numerical simulation by generating 10,000 values of the instantaneous amplitude, at the same sidereal time 5:30 of Pease's Mt. Wilson observation. As always, the CMB kinematical parameters were used to bound the random Fourier components of the stochastic velocity field Eqs.(6.64) and (6.65). The resulting histogram is reported in Fig.6.5.

From this histogram one obtains a mean simulated amplitude $\langle A_2^{\text{simul}} \rangle \sim$ 0.014. This corresponds to replace the value of the scalar velocity $\tilde{v}(t) \sim 369$ km/s Eq.(6.53), at the sidereal time of Pease's observation, in the relation for the statistical average of the amplitude

$$\langle A_2(t) \rangle_{\text{stat}} = \frac{2\epsilon D}{\lambda} \frac{\langle v_x^2(t) + v_y^2(t) \rangle_{\text{stat}}}{c^2} \sim 9.0 \cdot 10^{-3} \frac{\tilde{v}^2(t)}{(300 \text{ km/s})^2} \qquad (6.87)$$

In the above relation we have replaced $D/\lambda \sim 2.9 \cdot 10^7$ (for optical path of fifty-five feet), $\epsilon \sim 2.8 \cdot 10^{-4}$ and taken into account the factor $\pi^2/18$ in Eq.(6.69) to relate the statistical average of the stochastic velocity field to its smooth counterpart $\tilde{v}^2(t)$.

Notice that the median of the amplitude distribution is about 0.007. As a consequence, the value $A_2 \sim 0.006$ lies well within the 70% Confidence Limit. Also, the probability content becomes large at very small amplitudes[7] and there is a long tail extending up to $A_2 \sim 0.030$ or even larger values.

The wide interval of amplitudes corresponding to the 70% C. L. (which in terms of the mean could also be expressed as $0.014^{+0.015}_{-0.012}$) indicates that, in our stochastic model, one could easily explain individual MPP observations with an amplitude as 0.002 or as 0.030 which is fifteen times larger. This is another crucial difference with a deterministic model of the ether-drift. In this traditional view, in fact, the amplitude can vary at most by a factor $r = (v_{\text{max}}/v_{\text{min}})^2$ where v_{max} and v_{min} are respectively the maximum and minimum daily projection of the earth velocity in the interferometer plane. Therefore, since for the known types of cosmic motion r varies typically by a factor of two, the observation of such large fluctuations in the data would induce to conclude, in a deterministic model, that there is some systematic effect which modifies the measurements in an uncontrolled way. Then, if the ether drift has such an irregular nature, it may become understandable the MPP reluctance to quote the individual results and instead report those particularly small combinations obtained from large samples of data after averaging and further subtractions.

We emphasize that this general picture of a highly irregular phenomenon, which is characteristic of our stochastic model, is in agreement with the mentioned strong fluctuations of the observable velocity found by

---

[7]Strictly speaking, for a more precise comparison with the data, one should fold the histogram with a smearing function which takes into account the finite resolution Δ of the apparatus. This smearing would force the curve to bend for $A_2 \to 0$ and tend to some limit which depends on Δ. Nevertheless, this refinement should not modify substantially the probability content around the median which is very close to $A_2 = 0.007$.

Miller in individual measurements and, as shown in the next section, it is also confirmed by our accurate reanalysis of Joos's very precise experiment.

6.9   Reanalysis of Joos's experiment

As anticipated in Chap.5, for the accuracy of the apparatus and of the data taking, Joos' experiment [22, 143] cannot be compared with the other experiments. As such, it deserves a more refined analysis and will play a central role to test the validity of our model of the ether drift. In this section we will report the results of the analysis first presented in ref. [38] by including, however, new refinements that were not given there.

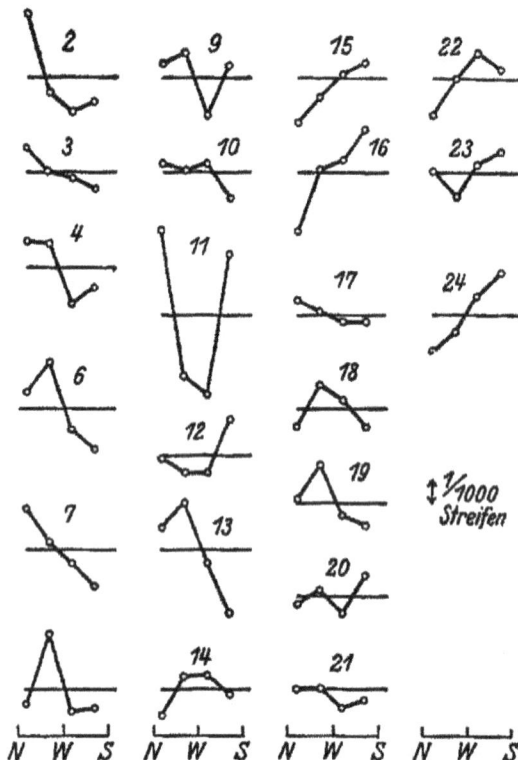

Fig. 6.6   *The selected set of data reported by Joos [22, 143]. The yardstick corresponds to 1/1000 of a wavelength so that the experimental dots have a size of about $0.4 \cdot 10^{-3}$. This corresponds to an uncertainty $\pm 0.2 \cdot 10^{-3}$ in the extraction of the fringe shifts.*

In Chap.5, we have mentioned that Joos's optical system was enclosed in a hermetic housing and that, traditionally, it was assumed that his measurements were performed in vacuum. While this aspect finds no definite confirmation in Joos' original articles, Swenson [98, 142] instead reports that fringe shifts were finally recorded with optical paths placed in a helium bath. Therefore, it seems safer to follow Swenson's explicit statement and assume that during the measurements Joos's interferometer was filled by gaseous helium at atmospheric pressure.

Observations were performed in Jena in 1930 starting at 2 P.M. of May 10th and ending at 1 P.M. of May 11th. Two measurements, the 1st and the 5th, were finally deleted by Joos with the motivation that there were spurious disturbances. The data were combined symmetrically, in order to eliminate the presence of odd harmonics, and the magnitude of the fringe shifts was typically of the order of a few thousandths of a wavelength. To this end, one can look at Fig.8 of [143] (reported here as our Fig.6.6) and compare with the shown size of $1/1000$ of a wavelength. From this picture, Joos decided to adopt $1/1000$ of a wavelength as an upper limit and deduced an observable velocity $v_{obs} \lesssim 1.5$ km/s. To derive this value, he used the fact that, for his apparatus, an observable velocity of 30 km/s would have produced a 2nd-harmonic amplitude of 0.375 wavelengths.

Though, since it is apparent that some fringe displacements were definitely larger than $1/1000$ of a wavelength, we have decided to perform a 2nd-harmonic fit to Joos's data and extract amplitude and phase from the 22 pictures, see Fig.6.7.

To better understand the nature of these data, we have also compared the average chi-square of the fit to Joos's fringe shifts with the average chi-square obtained from a 2nd-harmonic fit to many sets of 4 random numbers lying (as for Joos' data) in the range $[-4 \cdot 10^{-3}, +4 \cdot 10^{-3}]$. The average chi-square per degree of freedom obtained from Joos's data is about 0.8 while the corresponding average chi-square for the random sequences is about 8, i.e. ten times larger. This difference of an order of magnitude induces to give a physical meaning to the data in Fig.6.6.

In addition, a 2nd-harmonic fit to the substantial fringe shifts of Joos's picture 11 has an excellent chi-square, of comparable level and often better than the fits obtained from other observations with smaller values, see Fig.6.7. Therefore, there is no reason to delete the observation n.11.

Now, from the 2nd-harmonic fit the amplitudes A_2 can be extracted unambiguously. Instead, this is not true for the phases θ_0. The point is that it is not clear how to fix the reference angular values θ_k (with k=1,

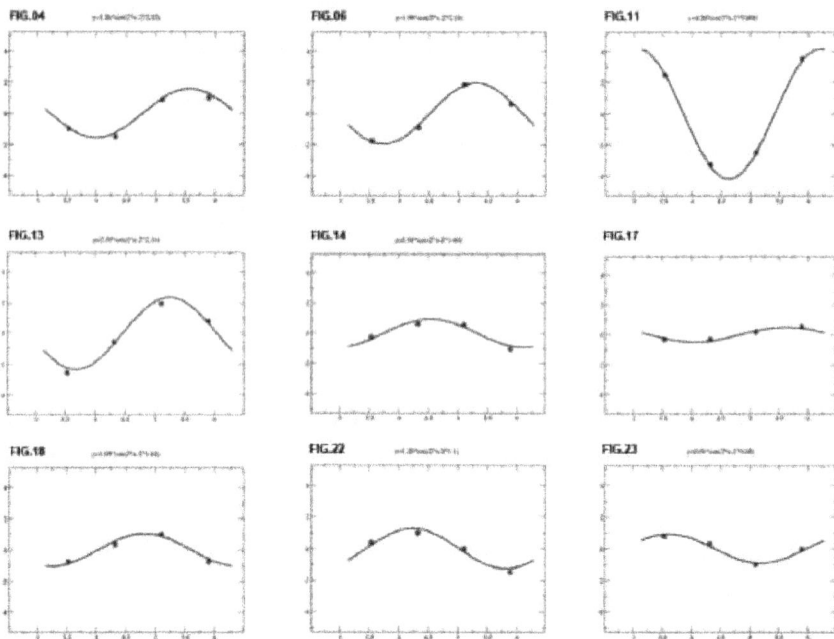

Fig. 6.7   *Some 2nd-harmonic fits to Joos's data. The y-axis gives the fringe shifts in units 10^{-3}. The x-axis gives the angles in radiants for a misalignment angle $\theta^* = 17$ degrees (see text).*

2, 3, 4) in Fig.6.6 for the fringe shifts. Thus, we could choose the set $\theta_k \equiv (360^\circ, 315^\circ, 270^\circ, 225^\circ)$ (consistently with our convention East= 90 degrees). However, in Fig.6.6 there is a small misalignment angle $\theta^* \sim 17^\circ$ (which actually from [22] might instead be 22.5°) between the dots of Joos's fringe shifts and the N, W, and S marks. Therefore, one could also adopt another set of values, namely $\theta_k \equiv (360^\circ - \theta^*, 315^\circ - \theta^*, 270^\circ - \theta^*, 225^\circ - \theta^*)$. In a 2nd-harmonic fit

$$\frac{\Delta\lambda(\theta_k)}{\lambda} = A_2 \cos 2(\theta_k - \theta_0) \qquad (6.88)$$

these options for the reference angles θ_k would give exactly the same amplitude A_2 but different choices for the phase θ_0. This basic ambiguity should also be added to the standard uncertainty in the phase that, due to the 2nd-harmonic nature of the measurements, could always be changed by adding ± 180 degrees. Therefore, since clearly there is only one correct choice for the angles θ_k, we have preferred not to quote theoretical uncer-

Table 6.1   *The 2nd-harmonic amplitude obtained from the 22 Joos pictures.   The uncertainty in the extraction of these values is about $\pm 0.2 \cdot 10^{-3}$ (the size of the dots in Fig.6.6).   The mean experimental amplitude over the 22 determinations is $\langle A_2^{\mathrm{EXP}} \rangle = 1.4 \cdot 10^{-3}$.*

Picture	$A_2^{\mathrm{EXP}}[10^{-3}]$
2	2.05
3	0.75
4	1.60
6	2.00
7	1.50
8	1.55
9	1.10
10	0.60
11	4.15
12	1.20
13	2.35
14	0.95
15	1.15
16	1.65
17	0.50
18	1.05
19	1.25
20	0.35
21	0.45
22	1.25
23	0.95
24	1.65

tainties on the phases and just concentrate on the amplitudes. Their values are reported in Table 6.1 and in Fig.6.8. Note that the lowest amplitude obtained from the fit to Joos's data, namely $A_2^{\mathrm{fit}} = 0.35 \cdot 10^{-3}$ from observation 20, can be taken as an estimate of the resolution of his readings. This is better by a factor of two than the accuracy of $1/1500$ of a fringe reported by Illingworth for his individual measurements.

By computing mean and variance of these amplitudes one finds

$$\langle A_2^{\mathrm{EXP}} \rangle = (1.4 \pm 0.8) \cdot 10^{-3} \tag{6.89}$$

and from $v_{\text{obs}} \sim 30$ km/s $\sqrt{\frac{A_2^{\text{EXP}}}{A_2^{\text{class}}}}$ a corresponding observable velocity

$$v_{\text{obs}} \sim 1.8^{+0.5}_{-0.6} \text{ km/s}. \tag{6.90}$$

Then, by correcting with the helium refractive index, where $\epsilon \sim 3.3 \cdot 10^{-5}$ at room temperature and atmospheric pressure, from Eq.(6.48) one would obtain a true kinematic velocity $v \sim 226^{+63}_{-75}$ km/s.

However, this is only a first partial view of Joos's experiment. In fact, we have compared Joos's amplitudes with theoretical models of cosmic motion. To this end, one has first to transform the civil times of Joos's measurements into sidereal times. For the longitude 11.60 degrees of Jena, one finds that Joos's observations correspond to a complete round in sidereal time in which the value $\tau = 0^o \equiv 360^o$ is very close to Joos's picture 20. Then, by using Eqs.(6.52) and (6.53), one can use this input and compare with theoretical predictions for the amplitude which, for the given latitude $\phi = 50.94$ degrees of Jena, depend on the right ascension α and the angular declination γ. To this end, it is convenient to first re-write the theoretical amplitude as

$$A_2(t) = \frac{2\epsilon D}{\lambda} \frac{v_x^2(t) + v_y^2(t)}{c^2} \sim 2.5 \cdot 10^{-3} \frac{v_x^2(t) + v_y^2(t)}{(300 \text{ km/s})^2} \tag{6.91}$$

where we have used the numerical relation for Joos's experiment $\frac{D}{\lambda} \frac{(30\text{km/s})^2}{c^2} \sim 0.375$ and replaced $\epsilon \sim 3.3 \cdot 10^{-5}$ for the helium refractive index.

As a first approximation, we have assumed a smooth velocity field and thus replaced $v_x(t) \sim \tilde{v}_x(t)$, $v_y(t) \sim \tilde{v}_y(t)$ by using Eq.(6.53) for the scalar combination $\tilde{v}(t) \equiv \sqrt{\tilde{v}_x^2(t) + \tilde{v}_y^2(t)}$. Then we have fitted the amplitude data of Table 6.1 to the smooth form

$$A_2^{\text{smooth}}(t) = \text{const} \cdot \sin^2 z(t) \tag{6.92}$$

where $\cos z(t)$ is defined in Eq. (6.52). The results of the fit[8]

$$\alpha(\text{fit} - \text{Joos}) = 168^o \pm 30^o \qquad \gamma(\text{fit} - \text{Joos}) = -13^o \pm 14^o \tag{6.93}$$

confirm that the earth motion with respect to the CMB, which has $\alpha(\text{CMB}) \sim 168^o$ and $\gamma(\text{CMB}) \sim -7^o$, could serve as a useful model to describe the ether-drift data.

---

[8]Actually, there is another degenerate minimum at $\alpha = 348^o \pm 30^o$ and $\gamma = 13^o \pm 14^o$ because $\sin^2 z(t)$ remains invariant under the simultaneous replacements $\alpha \to \alpha + 180^o$ and $\gamma \to -\gamma$. However, due to the close agreement with the CMB parameters we have concentrated on solution (6.93).

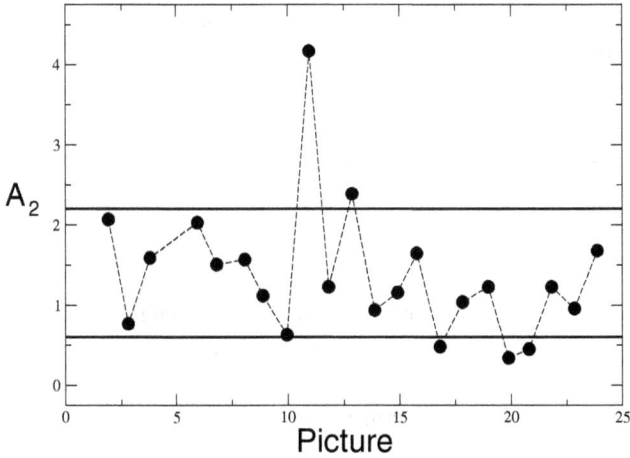

Picture

Fig. 6.8   *Joos's 2nd-harmonic amplitudes, in units* 10^{-3}. *The vertical band between the two lines corresponds to the range* $(1.4\pm0.8)\cdot10^{-3}$. *The figure is taken from ref. [38].*

However, in spite of the good agreement with the angular CMB $\alpha-$ and $\gamma-$values obtained from the fit Eq.(6.93), the nature of the strong fluctuations in Fig.6.8 remains unclear. Indeed, the amplitude of the observation 11 is more than ten times larger than the amplitudes from observations 20 and 21. This difference cannot be understood in a smooth model of the ether-drift.

Apart from this, there is also a sizeable difference in the absolute normalization of the average amplitude. In fact, by assuming the standard picture of smooth time modulations, the mean amplitude over all sidereal times can easily be obtained from the mean squared velocity Eq.(6.53)

$$\langle \tilde{v}^2(t) \rangle = V^2 \left(1 - \sin^2\gamma \sin^2\phi - \frac{1}{2}\cos^2\gamma\cos^2\phi \right). \qquad (6.94)$$

For the CMB and Jena, this gives $\sqrt{\langle \tilde{v}^2 \rangle} \sim 330$ km/s so that one would naively predict from Eq.(6.91)

$$\langle A_2^{\text{smooth}}(t) \rangle_{\text{day}} \sim 2.5 \cdot 10^{-3} \, \frac{\langle \tilde{v}^2(t) \rangle}{(300 \text{ km/s})^2} \sim 3.0 \cdot 10^{-3} \qquad (6.95)$$

to be compared with Joos's mean amplitude $\langle A_2^{\text{EXP}} \rangle = (1.4\pm0.8)\cdot10^{-3}$. In fact, in the standard picture, this experimental value leads to the previous estimate $\sqrt{\langle \tilde{v}^2 \rangle} \sim 226$ km/s and *not* to $\sqrt{\langle \tilde{v}^2 \rangle} \sim 330$ km/s. Therefore, it is necessary to change the theoretical model to try to make Joos's experiment completely consistent with the earth motion with respect to the CMB.

We have thus adopted the same model Eqs.(6.64), (6.65), to simulate stochastic variations of the velocity field and used the CMB kinematical parameters to fix the boundaries of the random variables. As discussed in the previous sections, in this model there will be a sizeable reduction of the amplitude as compared to its smooth prediction. In fact, by performing a full statistical average and using Eqs.(6.69) in Eq.(6.91), one finds

$$\langle A_2(t)\rangle_{\text{stat}} \sim 2.5 \cdot 10^{-3} \frac{\pi^2}{18} \frac{\tilde{v}^2(t)}{(300 \text{ km/s})^2} \sim 1.4 \cdot 10^{-3} \frac{\tilde{v}^2(t)}{(300 \text{ km/s})^2} \quad (6.96)$$

By also averaging over all sidereal times and replacing in Eq.(6.96) $\langle \tilde{v} \rangle \sim 330$ km/s, as for the CMB at Jena, we would now predict a mean amplitude of about $1.7 \cdot 10^{-3}$, rather than $3.0 \cdot 10^{-3}$, thus improving substantially the agreement with the experimental value $(1.4 \pm 0.8) \cdot 10^{-3}$.

After having understood these basic aspects, and after having fixed all theoretical inputs, we have started numerical simulations to study the dependence on the remaining parameters of the simulation, namely the number N of Fourier modes (in the available range $N \lesssim 10^7$) and the integer number s (the 'seed') which determines the random sequence. In particular, the dependence on the latter is usually quoted as theoretical uncertainty. For this reason, for the Piccard-Stahel experiment we had previously produced several simulations and taken into account the observed s-dependence of the results.

Here, we have started by doing something similar and first concentrated on the simplest statistical indicator, namely the mean amplitude $\langle A_2^{\text{simul}} \rangle$ obtained by averaging over all sidereal times. Quite in general, this can be evaluated for a variety of configurations which depend on the number n of measurements that one wants to simulate and the interval Δt between two consecutive measurements. For instance, Joos' experiment corresponds to $n = 24$ (actually $n = 22$ since Joos finally deleted two observations) and $\Delta t \sim 3600$ seconds. At the same time, the simulations become quite lengthy for large N, large n and small Δt. Therefore, we have first performed a scan of s-values for $N = 10^4$ and then studied a few s by increasing N. To give an idea of the spread of the central values, due to changes of the pair (N, s), we report below the approximate results of this analysis for some choices of the pair $(n, \Delta t)$

$$\langle A_2^{\text{simul}}(n = 24, \Delta t = 3600 \text{ s})\rangle \sim (1.7 \pm 0.8) \cdot 10^{-3} \quad (6.97)$$

$$\langle A_2^{\text{simul}}(n = 1440, \Delta t = 60 \text{ s})\rangle \sim (1.7 \pm 0.3) \cdot 10^{-3} \quad (6.98)$$

$$\langle A_2^{\text{simul}}(n = 240, \Delta t = 3600 \text{ s})\rangle \sim (1.8 \pm 0.5) \cdot 10^{-3} \quad (6.99)$$

Table 6.2 *The 2nd-harmonic amplitude obtained from a single simulation of 22 instantaneous measurements performed at Joos's times. The stochastic velocity components are controlled by the kinematical parameters* $(V, \alpha, \gamma)_{\text{CMB}}$ *as explained in the text. The mean amplitude over the 22 determinations is* $\langle A_2^{\text{simul}} \rangle = 1.38 \cdot 10^{-3}$.

Picture	$A_2^{\text{simul}}[10^{-3}]$
2	1.26
3	3.50
4	0.46
6	0.34
7	2.71
8	0.35
9	2.19
10	0.52
11	5.24
12	0.24
13	1.19
14	1.93
15	0.08
16	1.52
17	2.29
18	0.24
19	1.02
20	0.07
21	0.09
22	2.18
23	1.50
24	1.52

As it might be expected, the average $\langle A_2^{\text{simul}} \rangle$ becomes more stable by increasing the number of observations. Concerning the instantaneous values $A_2^{\text{simul}}(t_i)$, with $i = 1, .., n$, they have a large spread, about $(1 \div 5) \cdot 10^{-3}$ depending on the sidereal time. This is in agreement with the histogram we have obtained previously for the MPP experiment where the 70 % C.L. for the amplitude was found between 0.002 and 0.030. However this other spread can be reduced by starting to average the data in some interval of time t_0. In this case, the spread of the resulting average values $\langle A_2^{\text{simul}}(t_i) \rangle_{t_0}$ decreases as $\frac{1}{\sqrt{t_0}}$.

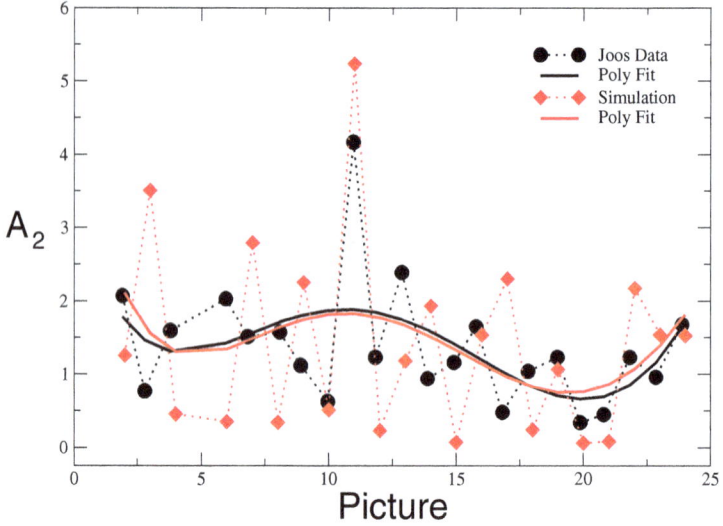

Fig. 6.9   *Joos' experimental amplitudes in Table 6.1 are compared with the single simulation of 22 measurements for fixed* (N, s) *in Table 6.2. By changing the pair* (N, s),
the typical variation of each simulated entry is $(1 \div 5) \cdot 10^{-3}$ *depending on the sidereal
time. We also show two 5th-order polynomial fits to the two different sets of values.
The figure is taken from ref. [38].*

We emphasize that, by performing extensive simulations, there are occasionally very large spikes of the amplitude at some sidereal times, of the
order $(10 \div 20) \cdot 10^{-3}$. The effect of these spikes gets smoothed when averaging over many configurations but their presence is characteristic of a
stochastic-ether model. With a standard attitude, where the ether drift is
a deterministic phenomenon expected to exhibit only smooth time modulations, the observation of such peaks would naturally be interpreted as a
spurious disturbance (Joos' deleted observations 1 and 5?).

After this preliminary study, we have then concentrated on the real
goal of our simulation, i.e. to compare with the *single* Joos configuration
of 22 entries in Table 6.1. To this end, one could first try to look for the
'best seed', or subset of seeds, which can minimize the difference between
the generated configurations and Joos' data. This standard task, usually
accomplished by minimizing a chi-square, is difficult to implement here.
In fact, it makes no sense to try to construct a function $\chi^2(s)$ and look
for its minima because a seed s and the closest seeds $s \pm 1$ give often

Table 6.3 *The 2nd-harmonic am-*
plitudes obtained by simulating the
averaging process over 10 hypothet-
ical measurements performed, at
each Joos' time, on 10 consecutive
days. The stochastic velocity com-
ponents are controlled by the kine-
matical parameters $(V, \alpha, \gamma)_{\text{CMB}}$ *as*
explained in the text. The effect of
varying the pair (N, s) *has been ap-*
proximated into a central value and
a symmetric error. The mean am-
plitude over the 22 determinations
is $\langle A_2^{\text{simul}} \rangle = 1.8 \cdot 10^{-3}$.

Picture	$A_2^{\text{simul}}[10^{-3}]$
2	2.5 ± 1.0
3	1.80 ± 0.85
4	1.95 ± 0.85
6	1.90 ± 0.85
7	1.65 ± 0.90
8	2.1 ± 1.0
9	2.0 ± 1.0
10	2.2 ± 1.2
11	2.4 ± 1.4
12	2.7 ± 1.6
13	2.3 ± 1.5
14	2.4 ± 1.4
15	1.85 ± 0.85
16	1.70 ± 0.75
17	1.20 ± 0.75
18	1.20 ± 0.70
19	1.15 ± 0.70
20	1.05 ± 0.70
21	1.25 ± 0.60
22	1.55 ± 0.60
23	1.60 ± 0.80
24	1.7 ± 1.0

vastly different configurations and chi-square. For this reason, we have followed an empirical procedure by forming a grid and selecting a set of seeds whose mean amplitude (for $n = 24$ and $\Delta t = 3600$ s) gets close to Joos's mean amplitude $\langle A_2^{\text{joos}} \rangle = 1.4 \cdot 10^{-3}$ for a large number N of Fourier modes. One of such seeds gave a sequence $\langle A_2^{\text{simul}} \rangle = 1.66, 1.40,$

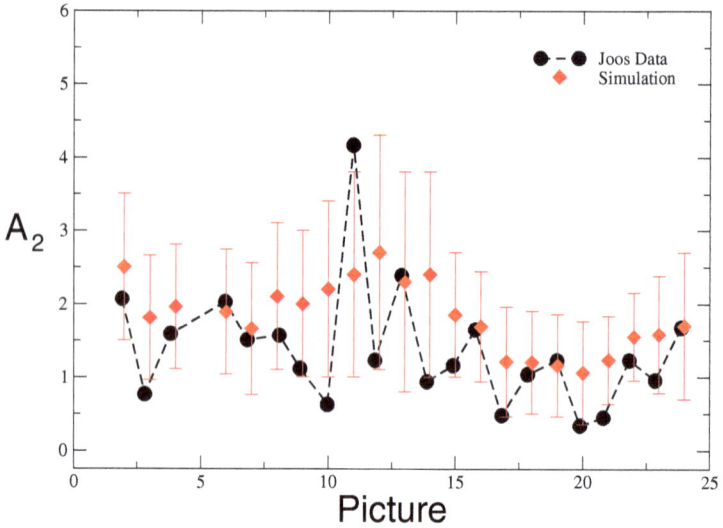

Fig. 6.10   *Joos' experimental amplitudes in Table 6.1 are compared with our simulation in Table 6.3. The figure is taken from ref. [38].*

1.08, 1.21 and 1.38 (in units 10^{-3}), for $N = 10^3$, 10^4, 10^5, 10^6 and $5 \cdot 10^6$ respectively, and the configuration with $N = 5 \cdot 10^6$ was finally chosen to give an idea of the agreement one can achieve between data and a single numerical simulation for fixed (N, s). The simulated values are reported in Table 6.2 and a graphical comparison with Joos' data is shown in Fig. 6.9. We emphasize that one should not compare each individual entry with the corresponding data since, by changing (N, s), the simulated instantaneous values vary typically of about $(1 \div 5) \cdot 10^{-3}$ depending on the sidereal time. Instead, one should compare the overall trend of data and simulation. To this end, we show two 5th-order polynomial fits to the two different sets of values.

A more conventional comparison with the data consists in quoting for the various 22 entries simulated average values and uncertainties. To this end, we have considered the mean amplitudes $\langle A_2^{\text{simul}}(t_i) \rangle$ defined by averaging, for each Joos' time t_i, over 10 hypothetical measurements performed on 10 consecutive days. For each t_i, the observed effect of varying (N, s) has been summarized into a central value and a symmetric error. The values are reported in Table 6.3 and the comparison with Joos' amplitudes is shown in Fig.6.10.

JOOS Sidereal Time for PICTURE 11

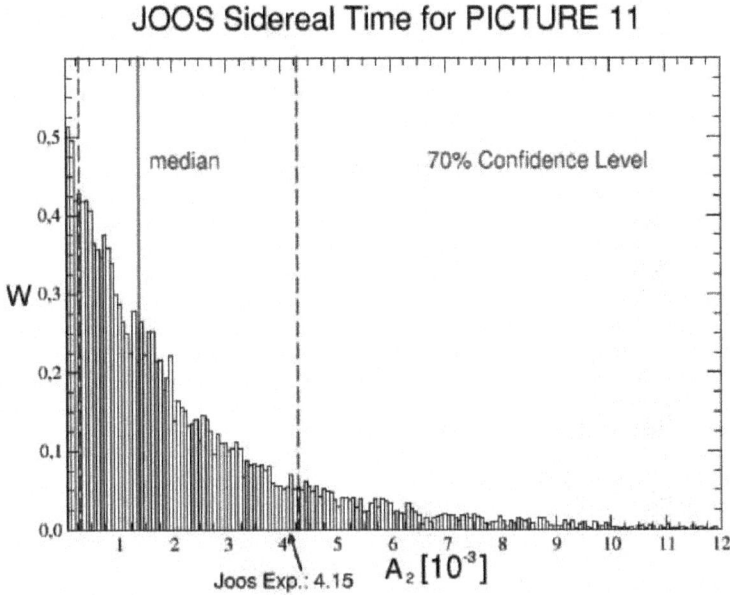

Fig. 6.11   *The histogram W of a simulation of 10,000 instantaneous amplitudes* $A_2(t)$ *a the same time of Joos's observation 11. The vertical normalization is to a unit area. We show the median, the 70% C.L. and Joos's experimental value.*

The spread of the various entries is larger at the sidereal times where the projection at Jena of the cosmic Earth's velocity becomes larger. The tendency of Joos' data to lie in the lower part of the numerical predictions in Table 6.3 mostly depends on our use of symmetric errors. In fact, by comparing in some case with the histograms of the basic generated configurations $A_2^{\text{simul}}(t_i)$, we have seen that our sampling method of $\langle A_2^{\text{simul}}(t_i)\rangle$, based on a grid of (N, s) values, typically underestimates the weight of the low-amplitude region in a prediction at the 70% C.L. . This can also be checked by considering the single simulation of Table 6.2 and counting the sizeable fraction of amplitudes $A_2^{\text{simul}}(t_i) \lesssim 0.5 \cdot 10^{-3}$. The more refined analysis based on the histograms is shown in Fig. 6.11 for Joos' picture 11 and in Fig.6.12 for picture 20.

In conclusion, after the first indication obtained from the fit Eq.(6.93), we believe that the link between Joos's data and the earth motion with respect to the CMB gets reinforced by our simulations. In fact, by inspection of the various tables and figures the experimental amplitudes and the characteristic scatter of the data are correctly reproduced. Thus, as already

JOOS Sidereal Time for PICTURE 20

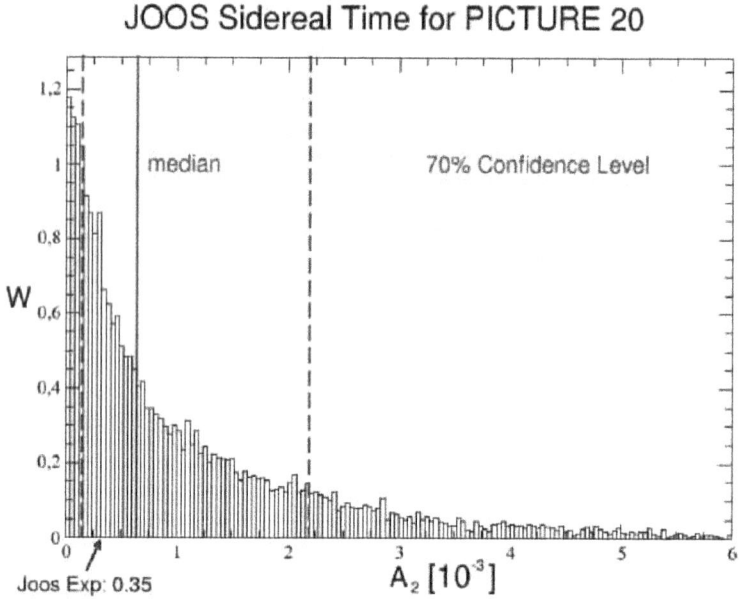

Fig. 6.12   *The histogram W of a simulation of 10,000 instantaneous amplitudes $A_2(t)$ a the same time of Joos's observation 20. The vertical normalization is to a unit area. We show the median, the 70% C.L. and Joos's experimental value.*

found for the Piccard-Stahel experiment, the agreement between the ether-drift data and the CMB parameters improves substantially by comparing with the stochastic model.

In particular, the previous value for the kinematical velocity $v \sim 226^{+63}_{-75}$ km/s, obtained by simply correcting with the helium refractive index the average observable velocity (6.90), has to be considerably increased if one allows for stochastic variations of the velocity field. In this case, by comparing with Eq.(6.96), Joos's experimental value $\langle A_2^{\mathrm{EXP}} \rangle = (1.4 \pm 0.8) \cdot 10^{-3}$ gives a range for the kinematical velocity

$$v \sim 305^{+85}_{-101} \text{ km/s} \qquad \text{(Joos)} \qquad (6.100)$$

6.10   Summary of the classical ether-drift experiments

In this final section we will summarize the results of our re-analysis of the classical ether-drift experiments in gaseous systems. As explained, this started from those theoretical arguments which suggest a picture of the physical vacuum as a stochastic medium, somehow similar to a turbulent

Table 6.4 *Our re-interpretation of the classical experiments in gaseous media. The analysis is based on the 2nd-harmonic experimental amplitudes extracted by assuming that the direction of the local drift is a completely random quantity with no definite limit by combining a large number of observations. The observable velocity, in km/s, is extracted from the ratio between these experimental amplitudes and the classical expectation through the relation $v_{obs} \sim 30$ km/s $\sqrt{A_2^{EXP}/A_2^{class}}$. The kinematical velocity $v[Eq.(6.48)]$, in km/s, is computed directly from the relation $v_{obs}^2 \sim 2\epsilon v^2$ Eq.(6.48). The second velocity $v[\text{stochastic}]$ is instead obtained from the statistical average Eq. (6.69) of the local velocity field. Numerically, it can be expressed as $\sqrt{18/\pi^2} \cdot v[Eq.(6.48)]$. The symbol $\pm...$ means that the experimental uncertainty cannot be estimated on the basis of the available information.*

Experiment	gas	v_{obs}	$v[Eq.(6.48)]$	$v[\text{stochastic}]$
Michelson(1881)	air	$19 \pm ...$	$800 \pm ...$	$1080 \pm ...$
Michelson-Morley(1887)	air	8.4 ± 1.7	355 ± 70	480 ± 95
Morley-Miller(1902-1905)	air	8.5 ± 1.5	359 ± 65	485 ± 85
Miller(1921-1926)	air	8.4 ± 2.2	355 ± 90	480 ± 120
Tomaschek (1924)	air	$7.7^{+2.1}_{-2.8}$	325^{+87}_{-116}	439^{+117}_{-157}
Kennedy(1926)	helium	< 5	< 600	< 810
Illingworth(1927)	helium	$2.4^{+0.8}_{-1.2}$	295^{+98}_{-148}	398^{+132}_{-200}
Piccard-Stahel(1926-1927)	air	$6.3^{+1.5}_{-2.0}$	266^{+62}_{-84}	359^{+84}_{-113}
Michelson-Pease-Pearson(1929)	air	$4.3 \pm ...$	$182 \pm ...$	$245 \pm ...$
Joos(1930)	helium	$1.8^{+0.5}_{-0.6}$	226^{+63}_{-75}	305^{+85}_{-101}

fluid which becomes statistically isotropic at small scales. In this perspective, the direction of the local drift in the plane of the interferometer becomes a completely random quantity which has no definite limit by combining a large number of measurements. Instead, one should concentrate on the 2nd-harmonic amplitudes A_2^{EXP} obtained from the individual measurements. These are positive-definite quantities and remain non-zero after any averaging procedure. By comparing with the classically expected value A_2^{class} for the traditional orbital value of 30 km/s, one can thus extract an observable velocity through the relation

$$v_{obs} \sim 30 \text{ km/s } \sqrt{\frac{A_2^{EXP}}{A_2^{class}}}. \tag{6.101}$$

From Eq.(6.48), one can then obtain a kinematical velocity v through the relation $v_{obs}^2 \sim 2\epsilon v^2$ with $\epsilon \equiv (\mathcal{N} - 1)$, $\mathcal{N}$ being the refractive index of the gaseous system in the optical paths ($\epsilon \sim 2.8 \cdot 10^{-4}$ or $\epsilon \sim 3.3 \cdot 10^{-5}$ respectively for air or gaseous helium at room temperature and atmospheric pressure).

As explained, however, this is only a first estimate. In fact, allowing for stochastic variations of the velocity field and taking their statistical average Eq.(6.69), one finds

$$\langle A_2(t)\rangle_{\text{stat}} \sim \frac{2\epsilon D}{\lambda} \frac{\langle v_x^2(t) + v_y^2(t)\rangle_{\text{stat}}}{c^2} \sim \frac{2\epsilon D}{\lambda} \frac{\pi^2}{18} \frac{\tilde{v}^2(t)}{c^2} \qquad (6.102)$$

where D= optical path, λ= light wavelength and $\tilde{v}(t)$ Eq.(6.53) fixes the boundaries of the random velocity components as in Eq.(6.68). By comparing with the analogous expression in a smooth, deterministic model of the drift

$$A_2^{\text{smooth}}(t) \sim \frac{2\epsilon D}{\lambda} \frac{\tilde{v}^2(t)}{c^2} \qquad (6.103)$$

it is evident that, from the same data, one would now get a velocity which is larger by a factor $\sqrt{18/\pi^2} \sim 1.35$. In this way, the agreement between the CMB parameters and the most precise Piccard-Stahel and Joos experiments improves considerably. Therefore, we also report the values of the kinematical velocity obtained in this other scheme. The summary of all determinations is shown in Table 6.4. The symbol $\pm$... means that, in some cases, the experimental uncertainty cannot be estimated on the basis of the available information.

6.11   A modern experiment with He-Ne lasers

The overall consistency of the kinematical velocities obtained from the various classical experiments in gaseous system and their substantial agreement with the value of 370 km/s obtained from the direct CMB observations is an important evidence which requires a check with the modern interference experiments with optical resonators. The essential ingredient is that the optical resonators should be filled by gaseous media. In this way, one could reproduce the experimental conditions of those old measurements with today's much greater accuracy. Such "non-vacuum" experiments would be along the lines of ref. [53] where just the use of optical cavities filled with different forms of matter was suggested as a useful complementary tool to investigate the possible existence of a preferred reference frame. The only delicate aspect concerns the high relative stability in temperature and pressure of the two optical cavities which is required to prevent possible spurious sources of the frequency shifts. However, with present technology and technical skill this should not represent a too serious problem. For instance, an important element to increase the overall stability and minimize systematic effects may consist in obtaining the two cavities from the same block of material as with the crossed optical cavity of ref. [168].

In units of their natural frequency ν_0, we then predict an instantaneous frequency shift between two orthogonal resonators

$$\left[\frac{\Delta\nu(\theta)}{\nu_0}\right]_{\text{gas}} = \left[\frac{\Delta\bar{c}_\theta}{c}\right]_{\text{gas}} \sim (\mathcal{N}_{\text{gas}} - 1)\,(v^2/c^2)\,\cos(2\theta) \qquad (6.104)$$

where θ is the angle between one of the two resonators and the laboratory velocity with respect to a hypothetical preferred reference frame. Then, for $v \sim 370$ km/s, the shift should be at the level $4 \cdot 10^{-10}$ or $5 \cdot 10^{-11}$ respectively for air or helium at atmospheric pressure. As such, it should be larger by orders of magnitude than the corresponding shift observed with vacuum resonators. Indeed, as we will see in the following chapter, the presently measured instantaneous value is $(\Delta\nu/\nu_0)_{\text{vacuum}} \sim 10^{-15}$.

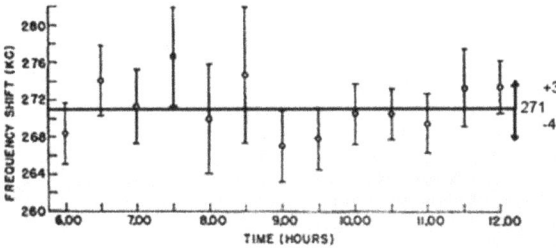

Fig. 6.13   *The frequency shifts of ref. [169] . The double arrow indicates the variation, with respect to the constant value, expected in the same model Eq.(6.104) used for the classical experiments. To this end, we have assumed the range $v = 315^{+20}_{-25}$ km/s which corresponds to the range of cosmic earth velocity for the latitude of Boston, at the time of the observations. In this way, once the average shift for 315 km/s is hidden in the much larger spurious frequency shift of 270 kHz, we would expect relative variations of about +3 and −4 kHz.*

This substantial enhancement is confirmed by the only modern experiment performed in similar conditions: the 1963 MIT experiment by Jaseja et. al [169] . Actually, at that time, they did not use optical resonators but were comparing directly the frequencies of two He-Ne lasers under 90 degrees rotations of the apparatus. However, the light from the lasers emerges from a He-Ne gas mixture and thus the two laser frequencies represent a measure of the two-way velocity of light, along orthogonal directions, in that environment. Finally, for a proper comparison, one has to subtract preliminarily a large systematic constant shift of about 270 kHz interpreted as being due to magnetostriction. As suggested by the same authors, this spurious effect, that was only affecting the overall normalization of the experimental $\Delta\nu$, can be subtracted by looking at the time variations of the data.

Now assuming a preferred frame, the shift is maximal for 90 degree rotations with respect to the true direction of the earth motion. This means a frequency shift

$$\text{Shift} \lesssim |\Delta\nu(\theta = 0) - \Delta\nu(\theta = \pi/2)| \sim 2(\mathcal{N}_{\text{He}-\text{Ne}} - 1) \ (v^2/c^2) \ \nu_0. \quad (6.105)$$

Then for a laser frequency $\nu_0 \sim 2.6 \cdot 10^{14}$ Hz, a refractive index $\mathcal{N}_{\text{He}-\text{Ne}} \sim 1.00004$ and a cosmic earth velocity of about 315^{+20}_{-25} km/s (as for the earth cosmic motion at the latitude of Boston and at the time of the observations) the expected frequency shift would be

$$\text{Shift} \lesssim 23 \ ^{+3}_{-4} \ \text{khz}. \quad (6.106)$$

Therefore, once the mean value is hidden in the much larger spurious shift of 270 kHz, we would expect typical relative variations of about $+3$ and -4 kHz (respectively above and below the mean value). This expectation is roughly consistent with residual variations of a few kHz shown in Fig.6.13. This fairly good agreement gives further motivations for the new series of experimental tests.

Chapter 7

Thus, Nonlocality is most naturally incorporated into a theory in which there is a special frame of reference. One possible candidate for this special frame of reference is the one in which the cosmic background radiation is isotropic. However, other than the fact that a realistic interpretation of quantum mechanics requires a preferred frame and the cosmic background radiation provides us with one, there is no readily apparent reason why the two should be linked.

L. HARDY, Physical Review Letters, 1992.

7.1 Modern experiments with vacuum optical resonators

As anticipated, in modern ether-drift experiments, the measured quantity is the shift $\Delta\nu$ between the frequencies of the light propagating inside two orthogonal optical cavities, see Fig.7.1.

The particular type of laser-cavity coupling used in the experiments is known in the literature as the Pound-Drever-Hall system [170, 171]. The details of this technique go beyond the scope of our book. However, the main ideas which are behind the mechanism are simple and beautifully explained in Black's tutorial article [172]. A laser beam is sent into a Fabry-Perot cavity which acts as a filter. Then, a part of the output of the cavity is fed back to the laser to suppress its frequency fluctuations. This method provide a very narrow bandwidth and has been crucial for the precision measurements we are going to describe.

145

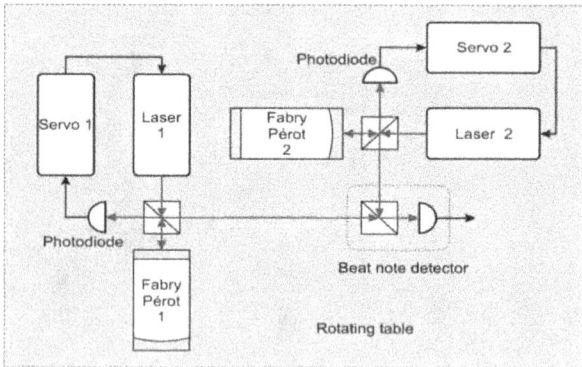

Fig. 7.1   *The scheme of a modern ether-drift experiment. The light frequencies are first stabilized by coupling the lasers to Fabry-Perot optical resonators. The frequencies ν_1 and ν_2 of the signals from the resonators are then compared in the beat note detector which provides the frequency shift $\Delta\nu = \nu_1 - \nu_2$. In present experiments a very high vacuum is maintained within the resonators.*

The first application to the ether-drift experiments was realized by Brillet and Hall in 1979 [25]. They were comparing the frequency of a CH_4 reference laser (fixed in the laboratory) with the frequency of a cavity-stabilized He-Ne laser ($\nu_0 \sim 8.8 \cdot 10^{13}$ Hz) placed on a rotating table. Since the stabilizing optical cavity was placed inside a vacuum envelope, the measured shift $\Delta\nu(\theta)$ was giving a measure of the anisotropy of the velocity of light in vacuum.

The stability of the cavity-laser system was found to be about $\pm$ 20 Hz for a 1-second measurement, and comparable to the stability of the reference CH_4 laser. It was also necessary to correct the data for a substantial linear drift of about 50 Hz/s.

By grouping the data in blocks of 10-20 rotations they found a signal with a typical amplitude of about 17 Hz (or a relative level 10^{-13}) and with a phase $\theta_0(t)$ which was randomly varying in the sidereal frame. Therefore, by increasing the statistics and projecting along the axis corresponding to the earth cosmic velocity obtained from the first CMB observations with U2 airplanes [173], the surviving average effect was substantially reduced down to about ± 1 Hz. Finally, by further averaging over a period of about 200 days, their final conclusion was that any residual ether-drift effect was a frequency shift $\Delta\nu = 0.13 \pm 0.22$ Hz, or at a fractional level $(1.5\pm2.5)\cdot 10^{-15}$.

With this standard averaging procedures a genuine ether drift is assumed to be a regular phenomenon depending deterministically on the earth velocity with respect to some fixed preferred frame. However, as emphasized in the previous chapter in connection with the classical experiments, the macroscopic earth motion (i.e. on a cosmic scale) could affect the microscopic propagation of light (inside an optical cavity) in a complicated, indirect way which depends ultimately on the nature of the physical vacuum. If, as suggested by some theoretical arguments, this has a fundamental stochastic nature, there might be irregular fluctuations of the local drift around the average earth motion. In this case, a genuine signal may easily be misinterpreted as spurious noise and averaging different observations becomes more delicate.

Since the 1979 Brillet-Hall article, substantial improvements have been introduced in the experiment. Just to have an idea, in present-day measurements [29, 168] with vacuum cavities the *instantaneous* fractional signal has been lowered from a few 10^{-13} to a few 10^{-15} and the linear drift from 50 Hz/s to about 0.05 Hz/s. The assumptions behind the analysis of the data, however, are basically unchanged. For this reason, by adopting the same strategy used for the classical experiments, we will first try to understand the magnitude of the irregular, instantaneous signal and then compare the data with numerical simulations performed within the simple analogy of a turbulent flow which becomes statistically homogeneous and isotropic at small scales. As explained in the previous chapter, in this approach the kinematical parameters entering the earth macroscopic motion are used to fix the boundary conditions for a microscopic velocity field which has an intrinsic non-deterministic nature.

To understand the magnitude of the instantaneous signal we have compared with Figure 9.a of ref. [168] and with Figure 4 of ref. [29]. These pictures confirm the idea of a highly irregular signal which varies approximately in the range ±1 Hz. For the adopted reference frequency $\nu_0 = 2.8 \cdot 10^{14}$ Hz, this is the mentioned fractional magnitude of a few 10^{-15}.

After having obtained this first information, we have tried to understand the meaning of this irregular signal, namely is it just spurious noise or could it represent a genuine ether drift? As a check, we have then compared with experiments where the optical cavities are made of different materials and are maintained at a cryogenic temperature, for instance the other experiments of ref. [174] and of ref. [28]. In these other two articles the instantaneous signal is not shown explicitly but it can be deduced from the typical variation of the signal which is observed over a characteristic time of $1 \div 2$ seconds. For a very irregular signal, in fact, this typical variation gives the magnitude of the signal itself and its value is again of a few 10^{-15}. Since it is unlike that spurious effects remain the same for experiments operating in so different conditions, we will explore the possibility that such an irregular fractional shift of a few 10^{-15} admits a physical interpretation.

Therefore, by applying to the physical vacuum the same model used successfully for the classical experiments, we will tentatively express this observed magnitude of the fractional shift in terms of the cosmic earth velocity and of a vacuum refractive index $\mathcal{N}_v$ as

$$\left. \frac{\Delta\nu(\theta)}{\nu_0} \right|_{\text{vacuum}} = \left. \frac{\Delta\bar{c}_\theta}{c} \right|_{\text{vacuum}} \sim (\mathcal{N}_v - 1) \, (v^2/c^2) \sim \mathcal{O}(10^{-15}). \qquad (7.1)$$

Then, for the typical value $v \sim 370$ km/s, we are lead to the concept of a refractive index for the physical vacuum (established in an apparatus placed on the earth surface) which differs from unity at the very small level 10^{-9}. In the following section we will explore the scenario suggested in ref. [57] which indeed could explain such a result.

7.2   An effective refractivity for the physical vacuum

The idea of an effective refractivity for the physical vacuum becomes natural by adopting a different view of the curvature effects observed in a gravitational field.

The usual perspective, derived from General Relativity, is that these effects require the introduction of a non-trivial metric field $g_{\mu\nu}(x)$ viewed as a fundamental modification of Minkowski space-time. By *fundamental*, we mean that deviations from flat space might also occur at extremely small scales, in principle comparable to the Planck length. Though, it is an experimental fact that many physical systems for which, at a fundamental level, space-time is exactly flat are nevertheless described by an effective curved metric in their hydrodynamic limit, i.e. at length scales that are much larger than the size of their elementary constituents.

For this reason several authors, see e.g. [145, 151, 152, 175], have started to explore those gravity-analogs (moving fluids, condensed matter systems with a refractive index, Bose-Einstein condensates,...) which are known in flat space. The ultimate goal is that, as with the deflection of light in Euclidean space when propagating in a medium of variable density, one might succeed to explain the curvature effects in a gravitational field in terms of the hydrodynamic excitations of an underlying form of (quantum) ether.

We believe that there is a value in this attempt. In fact, beyond the simple level of an analogy, there might be a deeper significance if the properties of the underlying medium could be matched with those of the physical vacuum of electroweak and strong interactions. In this case, the so called vacuum condensates, which play a crucial role for fundamental phenomena such as mass generation and quark confinement, could also represent a bridge between gravity and particle physics [153].

To be more definite, suppose that gravity originates from some long-range fields $s_k(x)$. By this we mean that their typical wavelengths are larger than some minimal scale (consistently with the experimental verifications [176] of the 1/r law) and that the deviation of the effective $g_{\mu\nu}(x)$ from the Minkowski tensor $\eta_{\mu\nu}$ can be expressed as

$$g_{\mu\nu}(x) - \eta_{\mu\nu} = \delta g_{\mu\nu}[s_k(x)] \tag{7.2}$$

with $\delta g_{\mu\nu}[s_k = 0] = 0$. In this type of approach, as in the original Yilmaz derivation [177], Einstein's equations for the metric should be considered as algebraic identities which follow directly from the equations of motion for the s_k's in flat space, after introducing a suitable stress tensor for these inducing-gravity fields[1].

In this way, one could (partially) fill the conceptual gap with classical General Relativity. As an immediate consequence, if the s_k's represent *excitations* of the physical vacuum, which therefore vanish identically in the equilibrium state, one could easily understand [145] why the huge condensation energy of the unperturbed vacuum plays no role, thus obtaining an intuitive solution of the cosmological-constant problem found in connection with the energy of the quantum vacuum [2].

This is not the place to discuss the various pros and cons of this type of approach. Instead, in our context of the ether-drift experiments, we will explore some possible phenomenological consequence. To this end, let us assume a zeroth-order model of gravity with a scalar field $s_0(x)$ which, at least on some coarse-grained scale, behaves as the Newtonian

---

[1] In the simplest, original Yilmaz approach [177] there is only one inducing-gravity field $s_0(x)$ which plays the role of the Newtonian potential. Introducing its stress tensor $t_\nu^\mu(s_0) = -\partial^\mu s_0 \partial_\nu s_0 + 1/2\delta_\nu^\mu \, \partial^\alpha s_0 \partial_\alpha s_0$, to match the Einstein tensor, produces differences from the Schwarzschild metric which are beyond the present experimental accuracy, see [178].

[2] In this sense, with this approach one is taking seriously Feynman's indication that "the first thing we should understand is how to formulate gravity so that it doesn't interact with the vacuum energy" [179].

potential. Then, how could its effects be effectively re-absorbed into a curved metric structure? At a pure kinematical level and regardless of the detailed dynamical mechanisms, the standard basic ingredients would be: (1) space-time dependent modifications of the physical clocks and rods and (2) space-time dependent modifications of the velocity of light. This point of view can be well represented by the following two citations:

Citation 1:
"It is possible, on the one hand, to postulate that the velocity of light is a universal constant, to define *natural* clocks and measuring rods as the standards by which space and time are to be judged and then to discover from measurement that space-time is *really* non-Euclidean. Alternatively, one can *define* space as Euclidean and time as the same everywhere, and discover (from exactly the same measurements) how the velocity of light and natural clocks, rods and particle inertias *really* behave in the neighborhood of large masses" [180].

Citation 2:
"Is space-time really curved? Isn't it conceivable that space-time is actually flat, but clocks and rulers with which we measure it, and which we regard as perfect, are actually rubbery? Might not even the most perfect of clocks slow down or speed up and the most perfect of rulers shrink or expand, as we move them from point to point and change their orientations? Would not such distortions of our clocks and rulers make a truly flat space-time appear to be curved? Yes." [181].

Therefore, within this interpretation of the space-time curvature, one might wonder about the fundamental assumption of General Relativity that, even in the presence of gravity, the velocity of light in vacuum c_γ is a universal constant, namely it remains the same, basic parameter c entering Lorentz transformations. Notice that, here, we are not considering the so called coordinate-dependent speed of light. Rather, our interest is focused on the value of the true, physical c_γ as, for instance, obtained from experimental measurements in vacuum optical cavities placed on the earth surface.

To understand the various aspects, a good reference is Cook's article "Physical time and physical space in general relativity" [182]. This article makes extremely clear which definitions of time and length, respectively $d\tau$ and dl, are needed if all observers have to measure the same, universal speed of light ("Einstein postulate"). For a static metric, these definitions are $d\tau^2 = g_{00}dt^2$ and $dl^2 = g_{ij}dx^i dx^j$. Thus, in General Relativity, the condition $ds^2 = 0$, which governs the propagation of light, can be expressed formally as

$$ds^2 = c^2 d\tau^2 - dl^2 = 0 \qquad (7.3)$$

and, by construction, yields always the same universal speed $c = dl/d\tau$.

For the same reason, however, if the physical units of time and space were instead defined to be $d\hat{\tau}$ and $d\hat{l}$ with, say, $d\tau = q\, d\hat{\tau}$ and $dl = p\, d\hat{l}$, the same condition

$$ds^2 = c^2 q^2 d\hat{\tau}^2 - p^2 d\hat{l}^2 = 0 \qquad (7.4)$$

would now be interpreted in terms of the different speed

$$c_\gamma = \frac{d\hat{l}}{d\hat{\tau}} = c\,\frac{q}{p} \equiv \frac{c}{\mathcal{N}_v}. \qquad (7.5)$$

The possibility of different standards for space-time measurements is thus a simple motivation for an effective vacuum refractive index $\mathcal{N}_v \neq 1$. As we are going to illustrate, this scenario can be tested and shown to be consistent with present ether-drift experiments.

For sake of clarity, we shall start our analysis from the unambiguous point of view of special relativity: the right space-time units are those for which the speed of light in the vacuum c_γ, when measured in an inertial frame, coincides with the basic parameter c entering Lorentz transformations. However, inertial frames are just an idealization. Therefore the appropriate realization is to assume *local* standards of distance and time such that the identification $c_\gamma = c$ holds as an asymptotic relation in the physical conditions which are as close as possible to an inertial frame, i.e. *in a freely falling frame* (at least by restricting light propagation to a space-time region small enough that tidal effects of the external gravitational potential $U_{\text{ext}}(x)$ can be ignored). This is essential to obtain an operational definition of the otherwise *unknown* parameter c.

With this premise, as originally discussed in ref. [57], light propagation for an observer S' sitting on the earth surface can be described with increasing degrees of accuracy starting from step (i), through (ii) and finally arriving to (iii):

(i) S' is considered a freely falling frame. This amounts to assume $c_\gamma = c$ so that, given two events which, in terms of the local space-time units of S', differ by (dx, dy, dz, dt), light propagation is described by the condition (ff='free-fall')

$$(ds^2)_{\text{ff}} = c^2 dt^2 - (dx^2 + dy^2 + dz^2) = 0 \qquad (7.6)$$

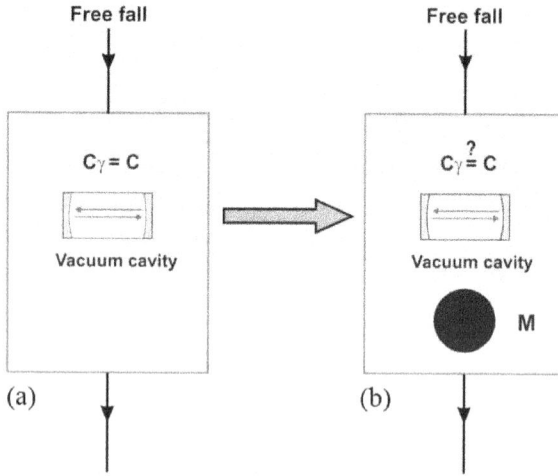

Fig. 7.2   *A pictorial representation of the effect of a heavy mass M carried on board of a freely-falling system, case (b). With respect to the ideal case (a), the mass M modifies the local space-time units and could introduce a tiny vacuum refractivity so that now $c_\gamma \neq c$.*

(ii) To a closer look, however, an observer S' placed on the earth surface can only be considered a freely-falling frame up to the presence of the earth gravitational field. Its inclusion can be estimated by considering S' as a freely-falling frame, in the same external gravitational field described by $U_{\text{ext}}(x)$, that however is also carrying on board a heavy object of mass M (the earth mass itself) which affects the local space-time structure, see Fig.7.2. To derive the required correction, let us denote by δU the extra Newtonian potential produced by the heavy mass M at the experimental set up where one wants to describe light propagation. Let us also denote by (dx, dy, dz, dt) the coordinate differences of the two chosen events (which for $M = 0$ coincide with the local space-time units of the freely-falling observer). According to General Relativity, and to first order in δU, light propagation for the S' observer is now described by

$$ds^2 = c^2 dt^2 (1 - 2\frac{|\delta U|}{c^2}) -$$

$$-(dx^2 + dy^2 + dz^2)(1 + 2\frac{|\delta U|}{c^2}) \equiv c^2 d\tau^2 - dl^2 = 0 \qquad (7.7)$$

where $d\tau^2 = (1 - 2\frac{|\delta U|}{c^2})dt^2$ and $dl^2 = (1 + 2\frac{|\delta U|}{c^2})(dx^2 + dy^2 + dz^2)$ are the physical units of General Relativity in terms of which one obtains the same universal value $dl/d\tau = c_\gamma = c$.

Though, to check experimentally the assumed identity $c_\gamma = c$ one should compare with a theoretical prediction for $(c - c_\gamma)$ and thus *necessarily* modify some formal ingredient of General Relativity. As a definite example, let us maintain the same definition of the unit of length but change the unit of time. The reason derives from the observation that physical units of time scale as inverse frequencies and that the measured frequencies $\hat\omega$ for $\delta U \neq 0$, when compared to the corresponding value ω for $\delta U = 0$, are red-shifted according to

$$\hat\omega = (1 - \frac{|\delta U|}{c^2})\, \omega. \tag{7.8}$$

Therefore, rather than the *natural* unit of time $d\tau = (1 - \frac{|\delta U|}{c^2})dt$ of General Relativity, one could consider the alternative, but natural (see our Citation 1), unit of time

$$d\hat t = (1 + \frac{|\delta U|}{c^2})\, dt. \tag{7.9}$$

Then, to reproduce Eq.(7.7), we can introduce a vacuum refractive index

$$\mathcal{N}_v \sim 1 + 2\frac{|\delta U|}{c^2} \tag{7.10}$$

so that the *same* Eq.(7.7) takes the form $(dl^2 \equiv (d\hat x^2 + d\hat y^2 + d\hat z^2))$

$$ds^2 = \frac{c^2 d\hat t^2}{\mathcal{N}_v^2} - (d\hat x^2 + d\hat y^2 + d\hat z^2) = 0. \tag{7.11}$$

This gives $dl/d\hat t = c_\gamma = \frac{c}{\mathcal{N}_v}$ and, for an observer placed on the earth surface, the relation

$$\frac{c - c_\gamma}{c} \sim \mathcal{N}_v - 1 \sim \frac{2G_N M}{c^2 R} \sim 1.4 \cdot 10^{-9}. \tag{7.12}$$

M and R being respectively the earth mass and radius.

Notice that, with this natural definition $d\hat t$, the vacuum refractive index associated with a Newtonian potential is the same usually reported in the literature, at least since Eddington's 1920 book [183], to explain in flat space the observed deflection of light in a gravitational field. The same expression is also suggested by the formal analogy of Maxwell equations in General Relativity with the electrodynamics of a macroscopic medium with dielectric function and magnetic permeability [184] $\epsilon_{ik} = \mu_{ik} = \sqrt{-g}\,\frac{(-g^{ik})}{g_{00}}$.

Indeed, in our case, from the relations $g^{il}g_{lk} = \delta_k^i$, $(-g^{ik}) \sim \delta_k^i \, g_{00}$, $\epsilon_{ik} = \mu_{ik} = \delta_k^i \mathcal{N}_v$, we obtain

$$\mathcal{N}_v \sim \sqrt{-g} \sim \sqrt{(1 - 2\frac{|\delta U|}{c^2})(1 + 2\frac{|\delta U|}{c^2})^3} \sim 1 + 2\frac{|\delta U|}{c^2}. \qquad (7.13)$$

A difference is found with Landau-Lifshitz' textbook [185] where the vacuum refractive index entering the constitutive relations is instead defined as $\mathcal{N}_v \sim \frac{1}{\sqrt{g_{00}}} \sim 1 + \frac{|\delta U|}{c^2}$. This alternative definition[3], if used to evaluate $(c - c_\gamma)$, corresponds to a different choice of the physical units and can also be taken into account as a theoretical uncertainty. We emphasize that this difference by a factor of 2 is not really essential. The main point is that c_γ, as measured in a vacuum cavity on the earth surface (panel **(b)** in our Fig.7.2), could differ at a fractional level 10^{-9} from the ideal value c, as operationally defined with the same apparatus in a true freely-falling frame (panel **(a)** in our Fig.7.2). In conclusion, this $c_\gamma - c$ difference can be conveniently expressed through a vacuum refractive index $\mathcal{N}_v$ in the form

$$\epsilon_v = \mathcal{N}_v - 1 \sim \frac{z}{2} \, 1.4 \cdot 10^{-9} \qquad (7.14)$$

where the factor $z/2$ (with $z= 1$ or 2) takes into account the mentioned theoretical uncertainty.

   (iii) Could one check experimentally if $\mathcal{N}_v \neq 1$? Today, the speed of light in vacuum is assumed to be a fixed number with no error, namely 299 792 458 m/s. Thus if, for instance, this estimate were taken to represent the value measured on the earth surface, in our picture and in an ideal freely-falling frame there should be a slight increase, namely $+\frac{z}{2}(0.42)$ m/s with $z = 1$ or 2. It seems hopeless to measure unambiguously such a difference because the uncertainty of the last precision measurements performed before the "exactness" assumption had precisely this order of magnitude, namely $\pm 4 \cdot 10^{-9}$ at the 3-sigma level or, equivalently, ± 1.2 m/s [188].

   Therefore, as pointed out in ref. [57], an experimental test cannot be obtained from the value of the average isotropic speed but, rather, from a possible *anisotropy* associated with a theoretical difference between c_γ and c. In fact, with a preferred frame, and if $\mathcal{N}_v \neq 1$, an isotropic light propagation as in Eq.(7.11) can only be valid for a special state of motion of the earth laboratory. This provides the definition of Σ while for a non-zero relative velocity **V** there are off diagonal elements $g_{0i} \neq 0$ in the

---

[3]A very complete set of references to these two possible alternatives for the vacuum refractive index in gravitational field is given by Broekaert [186], see his footnote [3].

effective metric [184]. The resulting two-way velocity would then be given by Eq.(6.42) with ϵ as in Eq.(7.14). On the basis of Eq.(6.44), and for the typical $v \sim 370$ km/s, we then expect a light anisotropy $\frac{|\Delta \bar{c}_\theta|}{c} \sim (\mathcal{N}_v - 1)(v/c)^2 \sim 10^{-15}$. As a matter of fact, this prediction is consistent with the presently most precise room-temperature vacuum experiment of ref. [29] and with the cryogenic vacuum experiment of ref. [28]. In particular, in the latter case this measured 10^{-15} level was about 100 times larger than the designed $O(10^{-17})$ short-term stability.

7.3 Simulations of experiments with vacuum optical resonators

Most recent ether-drift experiments measure the frequency shift $\Delta \nu$ of two *rotating* optical resonators. To this end, let us re-write Eq.(6.49) as

$$\frac{\Delta \nu(t)}{\nu_0} = \frac{\Delta \bar{c}_\theta(t)}{c} \sim \epsilon \frac{v^2(t)}{c^2} \cos 2(\omega_{\text{rot}}t - \theta_0(t)) \qquad (7.15)$$

where ω_{rot} is the rotation frequency of the apparatus. Therefore one finds

$$\frac{\Delta \nu(t)}{\nu_0} \sim 2S(t) \sin 2\omega_{\text{rot}}t + 2C(t) \cos 2\omega_{\text{rot}}t \qquad (7.16)$$

with $C(t)$ and $S(t)$ given in Eqs.(6.51).

To estimate the signal expected with vacuum optical resonators, we have performed [187] several numerical simulations in the isotropic stochastic model of Sec.6.5 with ϵ_v fixed as in Eq.(7.14) for $z = 2$. However, the theoretical uncertainty associated with the two possible choices $z = 1$ or 2 is also taken into account in the final formulas.

We first report in Fig.7.3 two typical sets for 2C(t) and 2S(t) during one rotation period $T_{\text{rot}} = 45$ seconds of the apparatus of ref. [26]. The two sets belong to the same random sequence and refer to two sidereal times that differ by 6 hours. The set $(V, \alpha, \gamma)_{\text{CMB}}$ was adopted to control the boundaries of the stochastic velocity components through Eqs.(6.52), (6.53) and (6.68). The value $\phi = 52$ degrees was also fixed to reproduce the average latitude of the laboratories in Berlin and Düsseldorf. For a laser frequency of $2.8 \cdot 10^{14}$ Hz [29], the interval $\pm 3 \cdot 10^{-15}$ of these dimensionless amplitudes corresponds to a random instantaneous frequency shift $\Delta \nu$ in the typical range ± 1 Hz. This is well consistent with the signal observed in ref. [29], see their Fig.4.

To compare with data extending over longer time intervals one has first to take into account the large, long-term drift which affects the experimental

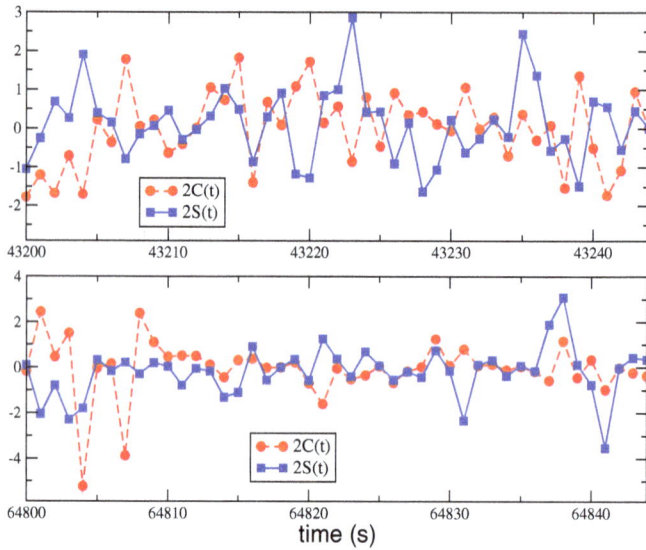

Fig. 7.3   *Two typical sets of 45 seconds for the instantaneous* $2C(t)$ *and* $2S(t)$ *in units* 10^{-15}. *The two sets belong to the same random sequence and refer to two sidereal times that differ by 6 hours. The boundaries of the stochastic velocity components in Eqs.(6.64) and (6.65) are controlled by* $(V, \alpha, \gamma)_{\mathrm{CMB}}$ *through Eqs.(6.53) and (6.68). For a laser frequency of* $2.8 \cdot 10^{14}$ *Hz [29], the interval* $\pm 3 \cdot 10^{-15}$ *corresponds to a typical frequency shift* $\Delta \nu$ *in the range* ± 1 *Hz. The figure is taken from ref. [187].*

frequency shift. For instance, for the presently most precise experiment of ref. [29], for time variations of several hours this drift is about ± 500 Hz, see their Fig.3 (top part). This is about 1000 times larger than the typical signal expected in our model, thus suggesting that we might be forced to abandon altogether the possibility of a precision test of our picture.

However, a way out derives from the observation that, although the frequency shift changes by such a large amount, still one can correct the data in order to achieve a much better stability. Indeed, by suitable modeling and subtraction of the drift, the typical variation of the shift over 1 second becomes about ± 0.24 Hz (see their Fig.3, bottom part) and thus at the level $\pm 8 \cdot 10^{-16}$. This means that, after correcting the data, the *local* properties of the signal, i.e. its characteristic variations over a time scale of 1 second, depend on the possible times t_i, t_j, t_k...of the observations to a negligible extent as compared to the original differences among the corresponding $\Delta \nu(t_i)$, $\Delta \nu(t_j)$, $\Delta \nu(t_k)$...Then, even if these were differing by a large amount, we can now get a test at the 10^{-15} level.

To compare with such a high short-term stability, we have thus simulated sequences of instantaneous measurements performed at regular steps of 1 second over an entire sidereal day. With such a type of simulation, we can also get an idea of the C_k and S_k, entering Eqs.(6.56) and (6.57), for a large but finite statistics (where one cannot get exactly zero as expected from Eqs.(6.69) and (6.70)). For a particular random sequence, the resulting histograms of $2C$ and $2S$ are reported in panels (a) and (b) of Fig.7.4.

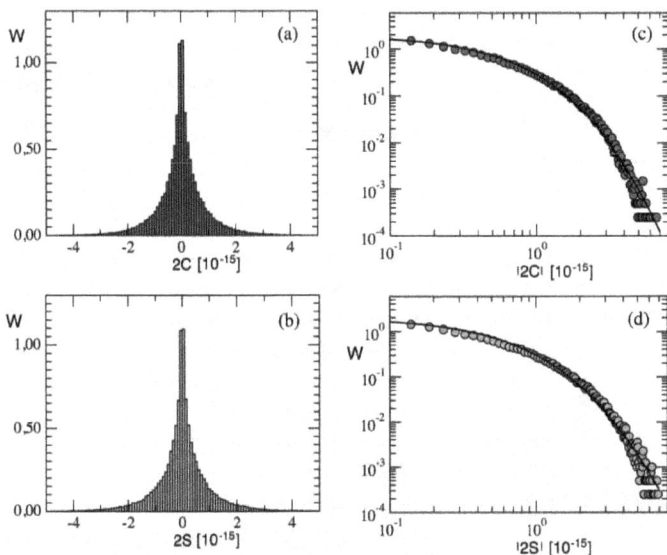

Fig. 7.4 *We show, see (a) and (b), the histograms W obtained from a single simulation of instantaneous measurements of $2C = 2C(t)$ and $2S = 2S(t)$ generated at regular steps of 1 second over an entire sidereal day. The vertical normalization is to a unit area. The mean values are $2C_0 = \langle 2C \rangle_{day} = -1.6 \cdot 10^{-18}$, $2S_0 = \langle 2S \rangle_{day} = 4.3 \cdot 10^{-18}$ and the standard deviations $\sigma(2C) = 8.7 \cdot 10^{-16}$, $\sigma(2S) = 9.6 \cdot 10^{-16}$. We also show, see (c) and (d), the corresponding plots in a log-log scale and the fits with Eq.(7.22). The boundaries of the stochastic velocity components in Eqs.(6.64) and (6.65) are controlled by $(V, \alpha, \gamma)_{CMB}$ through Eqs.(6.53) and (6.68). The figure is taken from ref. [187].*

In view of Eqs.(6.69) and (6.70) the non-zero averages $\langle 2C \rangle_{day} = 2C_0 = \mathcal{O}(10^{-18})$, $\langle 2S \rangle_{day} = 2S_0 = \mathcal{O}(10^{-18})$ should only be considered as statistical fluctuations around zero. The same holds true for the other C_k and S_k Fourier coefficients in Eqs.(6.56) and (6.57). By fitting the generated distributions to Eqs.(6.56) and (6.57) one gets values which are also $\mathcal{O}(10^{-18})$

or smaller and which fluctuate randomly around zero as expected. This simulated pattern is in complete agreement with the typical magnitude $\mathcal{O}(10^{-18})$ obtained in ref. [29] from the experimental data.

On the other hand, in such simulations of one-day measurements at steps of one second, the standard deviations around the 10^{-18} averages, say $\sigma_{\rm th}(2C)$ and $\sigma_{\rm th}(2S)$, are very stable ($z = 1$ or 2)

$$[\sigma_{\rm th}(2C)]_{1-\rm second} \sim \frac{z}{2}(8.7 \pm 0.8) \cdot 10^{-16} \qquad (7.17)$$

$$[\sigma_{\rm th}(2S)]_{1-\rm second} \sim \frac{z}{2}(9.6 \pm 0.9) \cdot 10^{-16}. \qquad (7.18)$$

Here the $\pm$ uncertainties reflect the observed variations due to the truncation of the Fourier modes in Eqs.(6.64), (6.65) and to the dependence on the random sequence.

From Eq.(7.16), by combining quadratically these two sigma's, we estimate

$$\left[\sigma_{\rm th}\left(\frac{\Delta\nu}{\nu_0}\right)\right]_{1-\rm second} \sim \frac{z}{2}(9 \pm 1) \cdot 10^{-16} \qquad (7.19)$$

so that, for a laser frequency $\nu_0 = 2.8 \cdot 10^{14}$ Hz [29] , we expect a typical spread

$$[\sigma_{\rm th}(\Delta\nu)]_{1-\rm second} \sim \frac{z}{2}(0.26 \pm 0.02) \text{ Hz} \qquad (7.20)$$

of the frequency shift measured every 1 second over a one-day period. For $z = 2$, this estimate is in very good agreement with the mentioned experimental value

$$[\sigma_{\rm exp}(\Delta\nu)]_{1-\rm second} \sim 0.24 \text{ Hz} \qquad (7.21)$$

which is reported in ref. [29] for an integration time of 1 second. Therefore, to the present best level of accuracy, this agreement strongly favours the value $z = 2$, which is the only free parameter of our scheme.

Our estimates are also well consistent with the analogous (but slightly less stringent) limit $\sigma_{\rm exp}(\frac{\Delta\nu}{\nu_0}) \sim 1.5 \cdot 10^{-15}$, at $1 \div 2$ seconds, placed by the cryogenic experiment of ref. [28] . Notice that this measured 10^{-15} level was about 100 times larger than the expected $\mathcal{O}(10^{-17})$ short-term stability. However, by the authors of ref. [28] , it was interpreted as a spurious effect due to a lack of rigidity of their cryostat. Probably, they have not considered the possibility of a genuine random signal and of intrinsic limitations placed by the vacuum structure.

We emphasize that the generated distributions are very different from a Gaussian shape, an aspect which is characteristic of probability distributions for instantaneous data in turbulent flows (see e.g. [62,63]) . To better appreciate the deviation from Gaussian behavior, in panels (c) and (d) we plot the same data in a log–log scale. The resulting distributions are well fitted by the so-called q–exponential function [65]

$$f_q(x) = a(1 - (1 - q)xb)^{1/(1-q)} \qquad (7.22)$$

with "entropic" index $q \sim 1.1$. This explains why, by performing extensive simulations, there might be occasionally large spikes of the instantaneous amplitude, up to $7 \cdot 10^{-15}$ or larger, when many Fourier modes sum up coherently (see the tails in panels (c) and (d) of Fig.7.4). The effect of these spikes, which lie at about 7 sigma's in terms of the standard deviations Eqs.(7.17) and (7.18), gets smoothed when averaging but their non-negligible presence (about 10^{-4} probability) is characteristic of the stochastic model. Otherwise, for a Gaussian distribution, 7 sigma's would correspond to a 10^{-11} probability.

As already observed for the classical experiments, another reliable indicator is the statistical average of the quadratic amplitude of the signal

$$A(t) \equiv 2\sqrt{S^2(t) + C^2(t)}$$

which is a positive-definite quantity and, as such, remains definitely non-zero after any averaging procedure. In this case, by using Eqs. (7.14) and (6.69), one finds ($z = 1$ or 2)

$$\langle A^{\mathrm{th}}(t) \rangle_{\mathrm{stat}} = \epsilon_v \frac{\tilde{v}^2(t)}{c^2} \frac{1}{3} \sum_{n=1}^{\infty} \frac{1}{n^2} = \frac{z}{2} (7.7 \cdot 10^{-16}) \frac{\tilde{v}^2(t)}{(300 \ \mathrm{km/s})^2}. \qquad (7.23)$$

By maintaining the CMB parameters $(V, \alpha, \gamma)_{\mathrm{CMB}}$ and fixing $\phi = 52$ degrees, one gets a daily average $\sqrt{\langle \tilde{v}^2 \rangle}_{\mathrm{day}} \sim 332 \ \mathrm{km/s}$ from the relation [57]

$$\langle \tilde{v}^2 \rangle_{\mathrm{day}} = V^2 \left(1 - \sin^2 \gamma \sin^2 \phi - \frac{1}{2} \cos^2 \gamma \cos^2 \phi \right). \qquad (7.24)$$

Thus, we predict a daily average amplitude ($z = 1$ or 2)

$$\langle A^{\mathrm{th}} \rangle_{\mathrm{day}} \sim \frac{z}{2} \ 9 \cdot 10^{-16} \qquad (7.25)$$

that, for a laser frequency $2.8 \cdot 10^{14}$ Hz, corresponds again to a typical instantaneous frequency shift $|\Delta\nu|_{\mathrm{th}} \sim \frac{z}{2} \ 0.26$ Hz. Other tests of the model will be possible if, besides the results of fits to the standard parameterizations Eqs. (6.56) and(6.57), also the basic instantaneous amplitudes $A(t)$,

$S(t)$ and $C(t)$ will become available. By comparing with these genuine data, we could also get other insights and improve on our simplest model of stochastic turbulence.

To conclude, we observe that a crucial test of our model consists in detecting tiny daily variations of the amplitude. This is a very difficult task due to the necessity of subtracting the mentioned systematic long-term drift which is much larger than the variation of a small fraction of Hz expected in our picture. Nevertheless, assuming that this subtraction could be done unambiguously to appreciate differences at the relative level 10^{-16}, for the CMB parameters at the latitude of Berlin-Düsseldorf, where the scalar velocity $\tilde{v}(t)$ in Eq.(6.53) changes in the range $260 \div 370$ km/s, from Eq.(7.23) we expect the typical range ($z = 1$ or 2)

$$\langle A^{\text{th}}(t)\rangle_{\text{stat}} = \frac{z}{2} (9 \pm 3) \cdot 10^{-16}. \tag{7.26}$$

More generally, if a daily variation of the amplitude will be detected, one could try to fit from the data the kinematical parameters (V, α, γ) entering Eq.(6.53).

7.4   Gaseous media vs. vacuum and solid dielectrics

Now, returning to the gas case, we will address in more detail the basic question of Chapt.1: independently of all symmetry arguments, why there should be a non-zero light anisotropy in the earth laboratory where (the container of) the gas is at rest? Moreover, only the region of refractive index infinitesimally close to the ideal vacuum $\mathcal{N} = 1$ has been analyzed. What about experiments performed in the other region where $\mathcal{N}$ differs substantially from unity, as in solid dielectrics?

Let us concentrate on the first question and let us summarize our results. We have found that by using the relation $\frac{|\Delta \bar{c}_\theta|}{c} \sim \epsilon(v^2/c^2)$ and correcting with the different refractive indexes, respectively $\epsilon \sim 2.8 \cdot 10^{-4}$ for air and $\epsilon \sim 3.3 \cdot 10^{-5}$ for gaseous helium, the same typical *kinematical* velocity $v \sim 300$ km/s can account for the observed effects. These values correspond to a light anisotropy $\frac{|\Delta \bar{c}_\theta|}{c} = \mathcal{O}(10^{-10})$ for air and to $\frac{|\Delta \bar{c}_\theta|}{c} = \mathcal{O}(10^{-11})$ for helium. Therefore since, for all practical purposes, a possible non-zero vacuum value $\frac{|\Delta \bar{c}_\theta|_v}{c} \lesssim 10^{-15}$ is irrelevant, the answer to our question requires to find the mechanism which *enhances* substantially the anisotropy in the gas case. To this end, it is natural to exploit the traditional *thermal* interpretation of the residuals of the classical experiments. This old argument, which gave the main motivation for Kennedy's replacement of air with gaseous helium in his optical paths, will now be illustrated by the explicit calculation of the gas refractive index $\mathcal{N}$.

The starting point is the Lorentz-Lorentz equation (see e.g. [190])

$$\frac{\mathcal{N}^2 - 1}{\mathcal{N}^2 + 3} = A_R\rho + B_R\rho^2 ... \tag{7.27}$$

where ρ is the molar density and $A_R = (4/3)\pi N_A\alpha$ is expressed in terms of the Avogadro number N_A and of the molecular polarizability α. The coefficient B_R takes into account two-body interactions and in our case, of air and helium at atmospheric pressure, this higher order term is completely negligible. Since $\mathcal{N}$ is very close to unity, we obtain the simplified formula for the gas refractivity

$$\epsilon = \mathcal{N} - 1 \sim \frac{3}{2}A_R\rho. \tag{7.28}$$

In the ideal-gas approximation, the molar density at Standard Temperature and Pressure (atmospheric pressure and zero centigrade or 273.15 K) has the well known value

$$\rho(STP) = \frac{P}{RT} = \frac{101325}{(8.314)(273.15)} \text{ mol} \cdot \text{m}^{-3} \sim$$
$$\sim 4.46 \cdot 10^{-5} \text{ mol} \cdot \text{cm}^{-3}. \tag{7.29}$$

Thus, for instance, for helium at STP and a wavelength $\lambda = 633$ nm, where $A_R \sim 0.52$ mol$^{-1} \cdot$ cm^3 [190], one finds $\epsilon \sim 3.5 \cdot 10^{-5}$.

The interesting aspect is that, in the ideal-gas approximation, the variation of the refractivity with the temperature has the very simple expression

$$-\frac{\partial\epsilon}{\partial T} \sim \frac{3}{2}A_R\frac{P}{RT^2} \sim \frac{\epsilon}{T}. \tag{7.30}$$

Therefore, by recalling the definition Eq.(6.17), a small temperature difference $\Delta T(\theta)$ induces a light anisotropy of typical magnitude

$$\frac{|\Delta\bar{c}_\theta|}{c} \sim |\bar{\mathcal{N}}(\theta) - \bar{\mathcal{N}}(\pi/2 + \theta)| \sim \frac{\epsilon|\Delta T(\theta)|}{T}. \tag{7.31}$$

We can thus obtain an experimental temperature difference from the 2nd-harmonic amplitudes A_2 in the fringe shifts

$$\frac{\Delta\lambda(\theta)}{\lambda} \sim \frac{2D}{\lambda}\frac{\Delta\bar{c}_\theta}{c} = A_2\cos 2\theta. \tag{7.32}$$

At room temperature, say $T = 288K + \Delta T$, this gives the relation

$$A_2^{\text{EXP}}(\text{air}) \sim \frac{2D}{\lambda} \cdot 10^{-9}\frac{\Delta T^{\text{EXP}}}{\text{mK}} \tag{7.33}$$

Table 7.1   *The average 2nd-harmonic amplitude observed in various classical ether-drift experiments and the resulting temperature difference obtained from Eqs.(7.33) and (7.34). The table is taken from ref. [187].*

Experiment	Gas	A_2^{EXP}	$\frac{2D}{\lambda}$	ΔT^{EXP}(mK)
Michelson-Morley(1887)	air	$(1.6 \pm 0.6) \cdot 10^{-2}$	$4 \cdot 10^7$	0.40 ± 0.15
Miller(1925-1926)	air	$(4.4 \pm 2.2) \cdot 10^{-2}$	$1.12 \cdot 10^8$	0.39 ± 0.20
Illingworth(1927)	helium	$(2.2 \pm 1.7) \cdot 10^{-4}$	$7 \cdot 10^6$	0.29 ± 0.22
Tomaschek (1924)	air	$(1.0 \pm 0.6) \cdot 10^{-2}$	$3 \cdot 10^7$	0.33 ± 0.20
Piccard-Stahel(1928)	air	$(2.8 \pm 1.5) \cdot 10^{-3}$	$1.28 \cdot 10^7$	0.22 ± 0.12
Joos(1930)	helium	$(1.4 \pm 0.8) \cdot 10^{-3}$	$7.5 \cdot 10^7$	0.17 ± 0.10

and

$$A_2^{EXP}(\text{helium}) \sim \frac{2D}{\lambda} (1.1 \cdot 10^{-10}) \frac{\Delta T^{EXP}}{mK}. \qquad (7.34)$$

The experimental values from the various experiments are reported in Table 7.1.

Our calculation shows that the old estimates of $1 \div 2$ mK by Kennedy, Shankland and Joos (see [47]) were too large, by about one order of magnitude. At the same time, the six determinations in Table 3 are well consistent with each other as shown by the excellent chi-square $(2.4/(6 - 1) = 0.48)$ of their mean

$$\langle \Delta T^{EXP} \rangle = (0.26 \pm 0.06) \text{ mK}. \qquad (7.35)$$

Therefore, the small residuals of the classical experiments in gaseous media can also be understood as thermal effects of a *non-local* origin. This suggests a possible relation with the CMB temperature dipole of ± 3 mK or with the fundamental energy flow which, on the basis of general arguments, is expected in a Lorentz-non-invariant vacuum. While, at present, such explanations have no definite quantitative basis, this thermal interpretation is important to get a consistent picture of all ether-drift experiments.

In fact, armed with this thermal interpretation, we can then address the second question concerning the ether-drift experiments in solid dielectrics, of the type performed by Shamir and Fox [30] in 1969. They were aware that the Michelson-Morley experiment did not yield a strictly zero result: "The non-zero result might have been real and due to the fact that the experiment was performed in air and not in vacuum" [30]. Thus, with $\mathcal{N}$ values substantially above unity, and within the traditional Lorentz-contraction interpretation, one might expect to observe a large ether-drift

$\frac{|\Delta \bar{c}_\theta|}{c} \sim (\mathcal{N}'^2 - 1)\beta^2 \sim \beta^2 \sim 10^{-6}$. The search for such effect was the motivation for their experiment in perspex ($\mathcal{N} = 1.5$). Since no such enhancement was observed, they concluded that the experimental basis of special relativity was strengthened.

However, with a thermal interpretation of the residuals observed in gaseous media, the two different behaviors can be reconciled. In a strongly bound system as a solid, in fact, a small temperature gradient of a fraction of millikelvin would mainly dissipate by heat conduction without generating any appreciable particle motion or light anisotropy in the rest frame of the apparatus. Hence, the non-trivial, physical difference between experiments in gaseous systems and experiments in solid dielectrics. In the latter case, we do not expect any enhancement with respect to the pure vacuum case. This means that, with very precise measurements, a fundamental vacuum anisotropy $\frac{|\Delta \bar{c}_\theta|_v}{c} \lesssim 10^{-15}$ should also show up in strongly bound solid dielectrics.

To make this more explicit, let us re-express the two-way velocity in terms of the ratio between the θ−dependent refractive index $\bar{\mathcal{N}}(\theta) \equiv \frac{c}{\bar{c}_\gamma(\theta)}$ of Eq.(6.17) and its reference value $\mathcal{N}$ in the preferred Σ frame. For $\mathcal{N} = 1 + \epsilon$, this gives

$$\frac{\bar{\mathcal{N}}(\theta)}{\mathcal{N}} \sim 1 + (\mathcal{N} - 1)\beta^2 \left(2 - \sin^2 \theta\right). \tag{7.36}$$

This form is valid for the gas case but, in our picture, is also valid in the vacuum limit where it takes the form

$$\frac{\bar{\mathcal{N}}_v(\theta)}{\mathcal{N}_v} \sim 1 + (\mathcal{N}_v - 1)\beta^2 \left(2 - \sin^2 \theta\right) \tag{7.37}$$

with, see Eq.(7.14), $(\mathcal{N}_v - 1) \sim \frac{z}{2} 1.4 \cdot 10^{-9}$ and $z = 1$ or 2.

The existence of $\mathcal{N}_v$ produces a tiny difference between the refractive index defined relatively to the ideal value $c_\gamma = c$ and the refractive index defined relatively to the physical isotropic value $c_\gamma = c/\mathcal{N}_v$ measured on the earth surface. The percentage difference between the two definitions is proportional to $\mathcal{N}_v - 1$ and, for all practical purposes, can be ignored. For instance, only for extremely low pressures one should take into account that the $\mathcal{N}_{gas} \to 1$ limit means actually $\mathcal{N}_{gas} \to \mathcal{N}_v$.

More significantly, all materials exhibit a background vacuum anisotropy proportional to $(\mathcal{N}_v - 1)\beta^2 \sim 10^{-15}$. In our interpretation, for light propagation in gases (at room temperature and atmospheric pressure) this is much smaller than the anisotropy induced by the non-local thermal gradient we have previously considered, namely

$$\frac{\bar{\mathcal{N}}_{\text{gas}}(\theta)}{\mathcal{N}_{\text{gas}}} \sim 1 + (a_{\text{thermal}} + a_v)\beta^2 \left(2 - \sin^2\theta\right) \qquad (7.38)$$

with $a_{\text{thermal}} \sim (\mathcal{N}_{\text{gas}} - \mathcal{N}_v)$ and $a_v \sim (\mathcal{N}_v - 1) \sim 10^{-9}$. Under normal conditions for the gas (recall that $a_{\text{thermal}} \sim 2.8 \cdot 10^{-4}$ or $a_{\text{thermal}} \sim 3.3 \cdot 10^{-5}$ respectively for air or helium at room temperature and atmospheric pressure) the a_v term is numerically irrelevant.

For solid dielectrics, on the other hand, where no enhancement is expected, one should keep track of the vacuum term. To this end, first replace the average isotropic value

$$\frac{c}{\mathcal{N}_{\text{solid}}} \rightarrow \frac{c}{\mathcal{N}_v \mathcal{N}_{\text{solid}}} \qquad (7.39)$$

and then use Eq.(7.37) to replace $\mathcal{N}_v$ in the denominator with $\bar{\mathcal{N}}_v(\theta)$ to take into account the motion of the laboratory relatively to the preferred Σ frame. This is equivalent to define a $\theta-$dependent refractive index for the solid dielectric

$$\frac{\bar{\mathcal{N}}_{\text{solid}}(\theta)}{\mathcal{N}_{\text{solid}}} \sim 1 + (\mathcal{N}_v - 1)\beta^2 \left(2 - \sin^2\theta\right) \qquad (7.40)$$

so that

$$[\bar{c}_\gamma(\theta)]_{\text{solid}} = \frac{c}{\mathcal{N}_{\text{solid}}(\theta)} = \frac{c}{\mathcal{N}_{\text{solid}}} \left[1 - (\mathcal{N}_v - 1)\beta^2 \left(2 - \sin^2\theta\right)\right] \qquad (7.41)$$

with an anisotropy

$$\frac{[\Delta\bar{c}_\theta]_{\text{solid}}}{[c/\mathcal{N}_{\text{solid}}]} \sim (\mathcal{N}_v - 1)\beta^2 \cos 2\theta \sim \frac{z}{2} \, 1.4 \cdot 10^{-9} \cdot 10^{-6} \cos 2\theta \lesssim 10^{-15}. \qquad (7.42)$$

Thus, for light propagation in solids, we would predict the same type of irregular signal discussed for pure vacuum and shown in our Fig.7.3. This expectation is consistent with the other cryogenic experiment by Nagel et al. [24]. In fact, most electromagnetic energy propagates in a dielectric with refractive index $\mathcal{N} \sim 3$ (at microwave frequencies) but the typical, *instantaneous* determination (see their Fig.3 b) is again $|\frac{\Delta\nu}{\nu_0}| \lesssim 10^{-15}$ as in the vacuum case (and as in the vacuum case it is about 1000 times larger than the average determination $|\langle\frac{\Delta\nu}{\nu_0}\rangle| \lesssim 10^{-18}$ obtained by combining a large number of measurements). For this reason the persistence, in vacuum and in solid dielectrics, of the irregular 10^{-15} signal should be definitely established. Instead, the present statistical averages, at the level $|\langle\frac{\Delta\nu}{\nu_0}\rangle| \lesssim 10^{-18}$, have no particular significance. With a stochastic signal, there is no problem in reaching the level $|\langle\frac{\Delta\nu}{\nu_0}\rangle| \lesssim 10^{-19}$, $|\langle\frac{\Delta\nu}{\nu_0}\rangle| \lesssim 10^{-20}$ ... by simply increasing the number of observations.

Finally, a complementary test could be performed by placing the vacuum (or solid dielectric) optical cavities on board of a satellite, as in the OPTIS proposal [191]. In this case where, even in a flat-space picture, the effective vacuum refractive index $\mathcal{N}_v$ for the freely-falling observer is exactly unity, the typical instantaneous frequency shift should be much smaller (by orders of magnitude) than the corresponding 10^{-15} value measured with the same interferometer on the earth surface.

7.5 Summary and conclusions

Our analysis of both classical and modern ether-drift experiments is now at the end. Our aim was to check the standard interpretation of these measurements (from Michelson-Morley until the most recent measurements with optical resonators) as "null results", i.e. typical instrumental effects in experiments with better and better systematics. This check is necessary because, by accepting the idea of a preferred frame and if the velocity of light c_γ propagating in the various interferometers does not coincide *exactly* with the basic parameter c entering Lorentz transformation nothing would prevent, in principle, to observe an ether drift. For this reason, it is natural to enquire to which extent the experiments performed so far have really given null results. By changing the theoretical model, small residual effects which apparently represent spurious instrumental artifacts could acquire a definite physical meaning with substantial implications for both physics and the history of science.

Indeed, detecting a definite non zero ether drift, which could be correlated with the earth cosmic motion deduced from the direct CMB observations with satellites in space, would definitely confirm that the reference frame, say Σ, where the dominant CMB dipole anisotropy vanishes exactly plays the role of preferred frame for relativity. Namely, as in the original Lorentzian formulation, Lorentz transformations would still remain exact to connect two observers in uniform translational motion but there would be a special frame of reference. The isotropy of the CMB radiation would then just *indicate* the existence of such a global Σ that we could decide to call "ether", but the cosmic microwave radiation itself would *not* coincide with this type of ether. Ultimate implications are far reaching. Think for instance of the possibility, with a preferred frame, to reconcile faster-than-light signals with causality and thus provide a very different view of the apparent non-local aspects of the quantum theory[4].

So far, the role of Σ as a preferred frame for relativity has never been seriously considered. However, as we have pointed out, this interpretation is not unexpected for at least two reasons. On the one hand, the observed CMB dipole can be reconstructed, to good approximation, from the various peculiar motions which are involved, namely the rotation of the solar system around the galactic center, the motion of the Milky Way around the center of the Local Group and the motion of the Local Group of galaxies in the direction of that large concentration of matter known as the Great Attractor. Then, once a vanishing CMB dipole is equivalent to switching-off all peculiar motions, one naturally arrives to the concept of a global frame of rest determined by the average distribution of matter in the universe. On the other hand, we have also recalled the present view of the vacuum of particle physics, i.e. the lowest energy state of the theory, as a condensate of quanta. These quanta macroscopically populate the same zero-3-momentum state and thus, by definition select a certain reference frame. This suggests that the mentioned global frame could also reflect a vacuum structure with some degree of substantiality and, in this sense, could characterize non-trivially the form of relativity which is physically realized in nature.

Now, the observed CMB dipole indicates a motion of the solar system with an average velocity of 370 km/s toward a point in the sky of right ascension 168 degrees and declination -7 degrees. Therefore, within some theoretical model, one could try to deduce the same type of kinematical parameters from the ether-drift experiments.

To this end, we have first presented a modern version of Maxwell's original calculation for the anisotropy of the two-way velocity of light. By using simple symmetry arguments, in the infinitesimal region of refractive index $\mathcal{N} = 1 + \epsilon$ a possible non-zero anisotropy should scale as $\frac{|\Delta \bar{c}_\theta|}{c} \sim \epsilon v^2/c^2$, see Eq.(6.44), v being the earth velocity with respect to the hypothetical Σ. Therefore, due to the strong suppression, with respect to the classical

---

[4]The importance of establishing a link between CMB and ether-drift experiments is better illustrated by quoting from ref. [34] where Hardy discusses the implications of the typical nonlocality of the quantum theory: "Thus, Nonlocality is most naturally incorporated into a theory in which there is a special frame of reference. One possible candidate for this special frame of reference is the one in which the cosmic background radiation is isotropic. However, other than the fact that a realistic interpretation of quantum mechanics requires a preferred frame and the cosmic background radiation provides us with one, there is no readily apparent reason why the two should be linked".

estimate $\frac{|\Delta \bar{c}_\theta|}{c} \sim v^2/(2c^2)$, the size of the small residuals observed in the classical experiments in gaseous media (Michelson-Morley, Miller, Illingworth, Joos,..) can become consistent with the typical value $v \sim 370$ km/s obtained from the direct observations of the CMB with aircrafts and satellites. The essential point is contained in the relation $v_{obs}^2 \sim 2\epsilon v^2$ which connects the *kinematical* velocity v to a much smaller *observable* velocity v_{obs} which determines the magnitude of the fringe shifts.

For the full consistency of this interpretation, however, a change of perspective is needed. Namely, the irregular character of the data requires that the local velocity field $v_\mu(t)$ which governs the instantaneous value of light anisotropy in the laboratory, and as such the fringe shifts in the old experiments or the frequency shifts in the modern experiments, should *not* be identified with the global velocity field $\tilde{v}_\mu(t)$ as directly fixed by the earth cosmic motion. Instead, from general arguments related to the idea of the vacuum as an underlying stochastic medium, we have proposed that the relation between these two quantities might be *indirect* and similar to what happens in turbulent flows. This means that the local $v_\mu(t)$ could fluctuate randomly while the global $\tilde{v}_\mu(t)$ would just fix its typical boundaries. Thus, if turbulence becomes homogeneous and isotropic at small scales, one has a definite model where a genuine instantaneous signal can well coexist with vanishing statistical averages for all vectorial quantities.

In this alternative picture, the direction of the local drift in the plane of the interferometer is a completely random quantity which has no definite limit by combining a large number of observations. Therefore, one should concentrate on the positive-definite quadratic amplitude of the signal and on its time modulations. In this case, by restricting to the amplitude, we have found a good consistency of the residuals of all classical experiments with the value of 370 km/s obtained from the direct observations of the CMB, see Table 6.4 at the end of Chap.6. In particular, by fitting the time dependence of the amplitudes extracted from Joos's very precise observations (data collected during all 24 hours to cover the full sidereal day and recorded automatically by photocamera), one even gets some information on the right ascension and angular declination, namely $\alpha(\text{fit} - \text{Joos}) = (168\pm30)$ degrees and $\gamma(\text{fit} - \text{Joos}) = (-13\pm14)$ degrees, to compare with the present values $\alpha(\text{CMB}) \sim 168$ degrees and $\gamma(\text{CMB}) \sim -7$ degrees.

Our alternative view should thus be checked with a new series of tests in which the optical resonators, which are coupled to the lasers, are filled by gaseous media. This would reproduce the physical conditions of those early measurements with today's much greater accuracy. At present, a first rough check can be obtained from the time variations of a few kHz observed in the only modern experiment performed in similar conditions, namely the 1963 MIT experiment by Jaseja et. al [169] with He-Ne lasers (see the discussion given at the end of Chap.6).

Waiting for these new experiments, we have compared our picture with the frequency shift detected in modern vacuum experiments. The point is that for the physical vacuum the ideal equality $\mathcal{N} = 1$ might not be exact. For instance, as proposed in [57], an effective refractivity $\epsilon_v \sim 10^{-9}$ could account for the difference between an apparatus in an ideal freely-falling frame and an apparatus placed on the earth surface. In this case, for a typical earth velocity of 370 km/s, we would expect a genuine, stochastic frequency shift $\frac{|\Delta\nu(t)|}{\nu_0} \sim \epsilon_v(v/c)^2 \sim 10^{-15}$ which coexists with vanishing statistical averages for all vectorial quantities, such as the C_k and S_k Fourier coefficients extracted from a standard temporal fit to the data with Eqs.(6.56) and (6.57).

Our numerical simulations indicate that this expectation is well consistent with the presently most precise room-temperature experiment of ref. [29] and with the cryogenic experiment of ref. [28] (which is only less precise by about a factor of 2). By itself, this substantial agreement between experiments with different systematics indicates that the observed signal might have a genuine physical component and not just originate from spurious noise in the spacers and the mirrors of the optical resonators, as assumed so far. In fact, the estimates of these contributions [189] are based on the fluctuation-dissipation theorem and thus there is no obvious reason that experiments operating at so different temperatures exhibit the same instrumental effects. The unexplained agreement with ref. [28] is particularly striking in view of the factor 100 which exists between observed signal 10^{-15} and designed short-term stability $\mathcal{O}(10^{-17})$. Tentatively, the authors of [28] interpreted this discrepancy as being due to a lack of rigidity of their cryostat but, probably, they have not considered the possibility of a genuine random signal and of intrinsic limitations placed by the vacuum structure. In this different perspective, the alternative interpretation proposed in [57], and implemented here, should also be taken into account.

This becomes even more true in view of the very good agreement obtained between the experimental value for the spread of the instantaneous signal found in ref. [29], namely $\sigma_{\exp}(\Delta\nu) \sim 0.24$ Hz, and our corresponding simulated value $\sigma_{\rm th}(\Delta\nu) \sim (0.26 \pm 0.02)$ Hz for that experiment, with $z = 2$ in Eq. (7.14).

The agreement we have obtained looks very promising and opens the possibility to reconstruct the CMB dipole with precise optical measurements performed within the earth laboratory and thus definitely clarify the fundamental issue of a preferred frame. To this end, however, real *data* (and not just the results of *fits*) should become available. In fact, our model, besides implying vanishing statistical averages for all vectorial quantities, in agreement with the observations, makes other definite predictions. For instance, precise time modulations of the quadratic amplitude of the signal and non-Gaussian (i.e. long-tail) distributions for the individual measurements. Although, at present, modern experiments give no information on these aspects, this idea of long tails finds definite support in the statistical analysis of Miller's extensive observations, see Fig.1 of the paper by Shankland et al. [47] reported here in Chapt.5 as our Fig.5.4.

Finally, in the last section, we have addressed the possible physical mechanism which enhances the signal in gaseous media, respectively $\frac{|\Delta \bar{c}_\theta|}{c} = \mathcal{O}(10^{-10})$ and $\frac{|\Delta \bar{c}_\theta|}{c} = \mathcal{O}(10^{-11})$ for air or helium at atmospheric pressure, relatively to the instantaneous vacuum value $\frac{|\Delta \bar{c}_\theta|}{c} \lesssim 10^{-15}$ found in modern experiments on the earth surface. For instance, one could imagine a suitable interaction of the incoming radiation with the medium to produce a different polarization in different directions. Any such mechanism, however, should act in both gaseous matter and solid dielectrics with the final result that light anisotropy should always increase with the refractivity of the medium, in contrast with the experimental evidence.

Therefore, if the effect observed in gases has to be specific of such weakly bound forms of matter, its natural interpretation is in terms of a *non local* temperature gradient associated with the earth motion. This shows up in all classical experiments, in agreement with the traditional thermal interpretation of the observed residuals. Only, its average magnitude $\langle \Delta T \rangle = (0.26 \pm 0.06)$ mK is somewhat smaller than the old estimates (about $1 \div 2$ mK) by Kennedy, Joos and Shankland. Conceivably, it might ultimately be related to the CMB temperature dipole of ± 3 mK or reflect the fundamental energy flow associated with a Lorentz-non-invariant vacuum state. While, at present, we have no definite quantitative insight, yet such thermal interpretation is important to understand the differences and the analogies among experiments in gaseous media, in vacuum and in solid dielectrics. Indeed, for experiments with optical cavities maintained in an extremely high vacuum (both at room temperature and in the cryogenic regime), where any residual gaseous matter is totally negligible, such tiny temperature variations cannot produce any observable effect.

On the other hand, in solid dielectrics a so small temperature gradient should mainly dissipate by heat conduction without generating any appreciable particle motion or light anisotropy in the rest frame of the apparatus. Hence, in solid dielectrics we do not expect any sizeable enhancement with respect to what is observed in the pure vacuum case. This expectation is consistent with the cryogenic experiment by Nagel et al. [24] where light propagates in a dielectric with refractive index $\mathcal{N} \sim 3$ (at microwave frequencies) but the typical, *instantaneous* determination (see their Fig.3 b) is again $\frac{|\Delta \bar{c}_\theta|}{c} \lesssim 10^{-15}$ as in the vacuum case (and as in the vacuum case is about 1000 times larger than the average determination $|\langle \frac{\Delta \bar{c}_\theta}{c} \rangle| \lesssim 10^{-18}$ obtained by combining a large number of measurements).

At present, our prediction of a fundamental irregular signal $\frac{|\Delta \bar{c}_\theta|}{c} \lesssim 10^{-15}$ is the only explanation for this observed agreement between so different experiments, namely ref. [29] with vacuum cavities at room temperature, vs. ref. [24] performed in a solid dielectric in the cryogenic regime. The definite persistence of such signal would confirm the existence of a fundamental preferred frame for relativity and would have substantial implications for our interpretation of non-locality in the quantum theory. Once definitely established, complementary tests should be performed by placing the vacuum (or solid dielectric) optical cavities on board of a satellite, as in the OPTIS proposal [191]. In this ideal free-fall environment, the typical instantaneous frequency shift should be much smaller (by orders of magnitude) than the corresponding 10^{-15} value measured with the same interferometers on the earth surface.

Bibliography

[1] A. A. Michelson and E. W. Morley, Am. J. Sci. **34** (1887) 333.

[2] J. C. Maxwell, Ether, Encyclopaedia Britannica, 9th Edition, 1878.

[3] R. J. Kennedy, Phys. Rev. **47** (1935) 965.

[4] M. Born, Einstein's Theory of Relativity, Dover Publ., New York, 1962.

[5] G. Holton, Isis **60** (1969) 132.

[6] W. M. Hicks, Phil. Mag. **3** (1902) 9.

[7] D. C. Miller, Rev. Mod. Phys. **5** (1933) 203.

[8] G. F. Fitzgerald, Science **13** (1889) 390.

[9] H. A. Lorentz, Michelson's Interference Experiment (in: The Principle of Relativity by H. A. Lorentz et al., Methuen 1923).

[10] H. A. Lorentz, Proc. Acad. Sciences Amsterdam, I (1899) 427.

[11] J. Larmor, Phil. Trans. Roy. Soc. **190** (1897) 205-300.

[12] J. Larmor, "Aether and Matter", Cambridge University Press, 1900.

[13] H. A. Lorentz, Proc. Acad. of Sci. Amsterdam, **6** 1904.

[14] H. Poincaré, C. R. Acad. Sci. Paris **140** (1905) 1504.

[15] E. Giannetto, Poincaré and the rise of special relativity, Hadronic Journal Supplement **10** (1995) 365.

[16] H. Poincaré, The Principles of Mathematical Physics, English Translation by G. B. Halsted, The Monist **XV**, 1-24, January 1905.

[17] A. Einstein, Ann. der Physik, 17 (1905) 891;(English translation in: The Principle of Relativity by H. A. Lorentz et al., Methuen 1923)

[18] A. Einstein, reported in the New York Times, January 16th, 1931, p.3.

[19] M. Born, Address to the Conference on 50 years of Relativity, Bern 1955, reprinted in Physics in my Generation, Pergamon Press, 1956.

[20] J. van Dongen, J. Arch. Hist. Exact Sci. **63** (2009) 655.

[21] K. K. Illingworth, Phys. Rev. **30** (1927) 692.

[22] G. Joos, Ann. d. Physik **7** (1930) 385.

[23] H. Müller et al., Appl. Phys. **B77** (2003)719.

[24] M. Nagel M. et al., Nature Comm. **6** (2015) 8174.

[25] A. Brillet and J. L. Hall, Phys. Rev. Lett. **42** (1979) 549.

[26] S. Herrmann, et al., Phys.Rev. D **80** (2009) 105011.

[27] Ch. Eisele, A. Newsky and S. Schiller, Phys. Rev. Lett. **103** (2009) 090401.

[28] M. Nagel et al., Ultra-stable Cryogenic Optical Resonators For Tests Of Fundamental Physics, arXiv:1308.5582[physics.optics].

[29] Q. Chen, E. Magoulakis, and S. Schiller,Phys. Rev. **D 93** (2016) 022003.

[30] J. Shamir and R. Fox, N. Cim. **B62**(1969)258.

[31] J. S. Bell, How to teach special relativity, in Speakable and unspeakable in quantum mechanics, Cambridge University Press 1987, p. 67.

[32] H. R. Brown and O. Pooley, The origin of the space-time metric: Bell's Lorentzian pedagogy and its significance in general relativity, in Physics meets Philosophy at the Planck Scale, C. Callender and N. Hugget Eds., Cambridge University Press 2000 (arXiv:gr-qc/9908048).

[33] H. R. Brown, Physical Relativity. Space-time structure from a dynamical perspective, Clarendon Press, Oxford 2005.

[34] L. Hardy, Phys. Rev. Lett. **68** (1992) 2981.

[35] See for instance, S. Liberati, S. Sonego and M. Visser, Ann. Phys. **298** (2002) 167.

[36] See for instance, V. Scarani et al., Phys. Lett. **A276** (2000) 1.

[37] H. A. Lorentz, The Theory of Electrons, Leipzig 1909, B. G. Teubner Editor.

[38] M. Consoli, C. Matheson and A. Pluchino, Eur. Phys. J. Plus **128** (2013) 71.

[39] V. Guerra and R. de Abreu, Eur. J. of Phys. **26** (2005) S117.

[40] M. Consoli, Found. of Phys. **45** (2015) 22.

[41] R. P. Feynman, R. B. Leighton and M. Sands, The Feynman Lectures on Physics, Addison Wesley Publ. Co. 1963.

[42] L. Onsager, Nuovo Cimento, Suppl. **6** (1949) 279.

[43] G. L. Eyink and K. R. Sreenivasan Rev. Mod. Phys. **78** (2006) 87.

[44] A. N. Kolmogorov, Dokl. Akad. Nauk SSSR **30** (1940) 4; English translation in Proc. R. Soc. **A 434** (1991) 9.

[45] M. Consoli, A. Pluchino, A. Rapisarda A. and S. Tudisco, Physica **A394**(2014) 61.

[46] A. A. Michelson, Am. J. Sci, **22** (1881) 120.

[47] R. S. Shankland et al., Rev. Mod. Phys.**27** (1955) 167.

[48] M. Consoli, A. Pluchino and A. Rapisarda, Europhysics Lett. **113** (2016) 19001.

[49] M. Yoon and D. Huterer, Ap. J. Lett. **813** (2015) L18.

[50] J. C. Mather, Rev. Mod. Phys. **79** (2007) 1331.

[51] G. F. Smoot, Rev. Mod. Phys. **79** (2007) 1349.

[52] G. Joos, Letters to the Editor, Phys. Rev. **45** (1934) 114.

[53] H. Müller, Phys. Rev. D **71** (2005) 045004.

[54] G. 't Hooft, Search of the Ultimate Building Blocks, Cambridge Univ. Press, 1997.

[55] M. Consoli and P.M. Stevenson, Int. J. Mod. Phys. **A15** (2000) 133.

[56] See, for instance, R. F. Streater and A. S. Wightman, PCT, Spin and Statistics, and all that, W. A. Benjamin, New York 1964.

[57] M. Consoli and L. Pappalardo, Gen. Rel. and Grav. **42** (2010) 2585.

[58] A. Loeb, Int. J. of Astrobiology **13** (2014) 337.

[59] G. Nicolis and I. Prigogine, Self-Organization in Non-Equilibrium Systems, Wiley-Interscience, New York 1971.

[60] G. Pruessner G., Self-Organised Criticality, Cambridge University Press, Cambridge, 2012.

[61] P. Allegrini, M. Giuntoli, P. Grigolini and B. J. West, Chaos, Solit. Fract., **20** (2004) 11.

[62] K. R. Sreenivasan, Rev. Mod. Phys. **71**, Centenary Volume 1999, S383.

[63] C. Beck, Phys. Rev. Lett. **98** (2007) 064502.

[64] C.Beck and E.G.D.Cohen, Physica A **322** (2003) 267.

[65] C.Tsallis, Introduction to Nonextensive Statistical Mechanics. Approaching a Complex World, Springer (2009).

[66] E. T. Whittaker, A History of the Theories of Aether and Electricity, Dover Publications, Inc. New York 1989.

[67] Conceptions of ether, Edited by G. N. Cantor and M. J. S. Hodge, Cambridge University Press 1981.

[68] R. Descartes, Principles of Philosophy II, 18, Italian Translation, Laterza, Bari 2000.

[69] *ibidem* II, 17.

[70] R. Descartes, Letter to Morin, see J. Bennett, Space and Subtle Matter in Descartes's Metaphysics, in New Essays on the Rationalists, R. Gennaro and C. Heunemann Eds., Oxford University Press 1999.

[71] J. W. Lynes, Journ. Hist. of Ideas **43** (1982) 55.

[72] R. Descartes, The World or Treatise on Light, 4, 8, Italian Translation, Laterza, Bari 2000.

[73] R. Descartes, Principles of Philosophy II, 33, Italian Translation, Laterza, Bari 2000.

[74] C. F. von Weizäcker, Grosse Physiker, Italian Translation, Donzelli Editore, Roma 2002.

[75] I. Newton, The Principia, Translated by A. Motte, Prometheus Books, Amherst NY, 1995.

[76] J. C. Maxwell, On Action at a Distance, The Scientific Papers of James Clerk Maxwell, Edited by W. D. Niven, Dover Publ. Inc., New York 1965, p. 311.

[77] J. Henry, Studies in History and Philosophy of Science **42** (2011) 11.

[78] I. Newton, Opticks: Or a Treatise of the Reflections, Refractions, Inflections and Colours of Light, Cosimo Inc.,2007.

[79] Philip E. B. Jourdain, The Monist **25** (1915) 418. Available at http://www.jstor.org/stable/27900547.

[80] P. M. Heimann, Ether and imponderables, contributed paper to Conceptions of ether, Edited by G. N. Cantor and M. J. S. Hodge, Cambridge University Press 1981, p.61.

[81] R. S. Westfall, Force in Newton's Physics: The Science of Dynamics in the Seventeenth Century, American Elsevier, New York 1971.

[82] B. Hall, Brit. J. Hist. Philosophy **14** (2006) 719.

[83] I. Kant, Methaphysical Foundations of Natural Science, Translated and Commented by J. Bennett, 2005. Available at www. earlymodern-texts.com/assets/pdfs/kant1786.pdf.

[84] I. Kant, Critique of pure reason, Translated and Edited by P. Guyer and A. W. Wood, Cambridge University Press 1998.

[85] I. Kant, Opus Postumum, Edited by E. Förster, Translated by E. Förster and M. Rosen, Cambridge University Press 1995.

[86] T. Young, Phil. Trans. R. Soc. Lond. 90 (1800) 106.

[87] The Bakerian Lecture. On the Theory of Light and Colours. By Thomas Young, M. D. F. R. S. Professor of Natural Philosophy in the Royal Institution 1802. JSTOR Digital Library.

[88] A. Fresnel, Oeuvres complètes, as quoted in ref. [66] .

[89] J. Stachel, Fresnel's (Dragging) Coefficient as a Challenge to 19th Century Optics of Moving Bodies, contribution to The Universe of General Relativity, Einstein Studies Vol. 11, Edited by A. J. Kox and J. Eisenstaedt, Birkhäuser 2005.

[90] O. V. Troshkin, Physica **A168** (1990) 881.

[91] H. E. Puthoff, Linearized turbulent flow as an analog model for linearized General Relativity, arXiv:0808.3401 [physics.gen-ph].

[92] T. D. Tsankov, Classical Electrodynamics and the Turbulent Aether Hypothesis, Preprint February 2009, unpublished.

[93] P. A. Davidson, Turbulence: An Introduction for Scientists and Engineers, Oxford University Press 2004.

[94] S. B. Pope, Turbulent Flows, Cambridge University Press 2000.

[95] M. J. Marcinkowski, Physica Status Solidi **152B** (1989) 9.

[96] A. M. Kosevic, The Crystal Lattice: Phonons, Solitons, Superlattices, Wiley-VCH Verlag Gmbh and Co. KGaA, Weinheim 2005.

[97] R. S. Shankland, Am. J. Phys. **32** (1964) 16.

[98] L. S. Swenson Jr., the Ethereal Aether, A History of the Michelson-Morley-Miller Aether-Drift Experiments, 1880-1930. University of Texas Press, Austin 1972.

[99] A. A. Michelson, Am. J. Sci. **22** (1881) 120.

[100] B. Haubold, H. J. Haubold, and L. Pyenson, Michelson's first ether-drift experiment in Berlin and Potsdam, in The Michelson Era in American Science, S. Goldberg and R. H. Steuwer Eds., AIP, New York 1988.

[101] A. A. Michelson and E. W. Morley, Am. J. Sci. **31** (1886) 377.

[102] A. A. Michelson, et al., Astrophys. Journ. **68** (1928) pag. 341-402.

[103] An incomplete list of references includes: W. Sutherland, Phil. Mag. **45** (1898) 23; A. Righi, N. Cimento **XVI** (1918) 213; *ibidem* **XIX** (1920) 141; *ibidem* **XXI** (1921) 187; G. Dalla Noce, N. Cimento **XXIV** (1922) 17. Righi's theory was also re-analyzed by P. Di Mauro, S. Notarrigo and A. Pagano, Quaderni di Storia della Fisica, **2** (1997) 101.

[104] M. Consoli and E. Costanzo, Phys. Lett. **A333** (2004) 355.

[105] M. Consoli and E. Costanzo, N. Cim. **119B** (2004) 393.

[106] M. A. Handshy, Am. J. of Phys. **50** (1982) 987.

[107] W. M. Hicks, Phil. Mag. **3** (1902) 256.

[108] C. Kittel, Am. J. Phys. **42** (1973) 726.

[109] M. N. Macrossan, Brit. J. Phil. Sci. **37** (1986) 232.

[110] R. de Abreu and V. Guerra, Electr. J. of Theor. Phys. **12** (2015) 183.

[111] A. Wiener, Lorentz Contraction: Explained at the Microscopic Level, M. S. thesis, The Blackett Laboratory, Imperial College London. 2009.

[112] A. Ungar, Found. of Phys. **30** (2000) 331.

[113] J. P. Costella et al., Am. J. Phys. **69** (2001) 837.

[114] K. O' Donnell and M. Visser, Eur. J. Phys. **32** (2011) 1033.

[115] E. Cassirer, Einstein's Theory of Relativity Considered from the Epistemological Point of View, The Monist **32** (1922), pp. 89-134.

[116] W. Pauli, Teoria della Relatività, Ed. Borighieri 1970.

[117] L. Kostro, Einstein and the ether, Italian translation, Ed. Dedalo, Bari 2001.

[118] B. Riemann, Sulle ipotesi che stanno alla base della geometria, a cura di R. Pettoello, Bollati Boringhieri, Torino 1999.

[119] U. Bottazzini and R. Tazzioli, Rev. Hist. Math. **1** (1995) 3.

[120] A. Einstein and L. Infeld, The Evolution of Physics, Cambridge University Press, 1938.

[121] E. W. Morley and D. C. Miller, Phil. Mag. **9** (1905) 680.

[122] R. Lalli, Ann. of Science **69** (2012) 153.

[123] K. R. Popper, The Logic of Scientific Discovery, Italian Translation by M. Trinchero, Einaudi 1970.

[124] D. C. Miller, Phys. Rev. **19** (1922) 407.

[125] D. C. Miller, Phys. Rev. **45** (1934) 114.

[126] R. Tomaschek, Ann. d. Physik, **73** (1924) 105.

[127] R. Tomaschek, Astron. Nachrichten, **219** (1923) 301, English translation.

[128] T. Roberts, An Explanation of Dayton Miller's Anomalous "Ether Drift" Result, arXiv:physics/0608238.

[129] F. James, MINUIT: Function minimization and error analysis, CERN Computing and Networks Division, Long Writeup D506, Geneva 1994.

[130] M. von Laue, Handbuch d. Experimentalphysik (1926), **XVII** 95.

[131] H. Thirring, Z. Physik **35** (1926) 723; Nature **118** (1926) 81.

[132] H. A. Múnera, APEIRON **5** (1998) 37.

[133] J. J. Nassau and P. M. Morse, Astrophys. Journ. **65** (1927) 73.

[134] D. C. Miller, Science **LXIII** (1926) 433.

[135] A. Piccard and E. Stahel, Compt. Rend. **183** (1926) 420; Naturwiss. **14** (1926) 935.

[136] A. Piccard and E. Stahel, Compt. Rend. **185** (1927) 1198; Naturwiss. **16** (1928) 25.

[137] A. Piccard and E. Stahel, Journ. de Physique et Le Radium **IX** (1928) No.2.

[138] SH. Cha, International Journal of Mathematical Models and Methods in Applied Sciences 1(4) (2007) 300.

[139] A. A. Michelson, F. G. Pease and F. Pearson, Nature **123** (1929) 88.

[140] A. A. Michelson, F. G. Pease and F. Pearson, J. Opt. Soc. Am. **18** (1929) 181.

[141] F. G. Pease, Publ. of the Astr. Soc. of the Pacific, **XLII** (1930) 197.

[142] Loyd S. Swenson Jr., Journ. for the History of Astronomy, **1** (1970) 56.

[143] G. Joos, Naturwiss. **38** (1931) 784.

[144] J. DeMeo, Dayton C. Miller Revisited, in Should the Laws of Gravitation be Reconsidered?, H. A. Múnera Ed., APEIRON Montreal 2011, pp. 285-315.
[145] G. E. Volovik, Phys. Rep. **351**, 195 (2001).
[146] M. Consoli, A. Pagano and L. Pappalardo, Phys. Lett. **A318** (2003) 292.
[147] M. Consoli and E. Costanzo, Eur. Phys. Journ. **C54** (2008) 285.
[148] M. Consoli and E. Costanzo, Eur. Phys. Journ. **C55** (2008) 469.
[149] Y. B. Zeldovich, Sov. Phys. Usp. **11** (1968) 381.
[150] S. Weinberg, Rev. Mod. Phys. **61** (1989) 1.
[151] C. Barcelo, S. Liberati and M. Visser, Class. Quantum Grav. **18** (2001) 3595.
[152] M. Visser, C. Barcelo and S. Liberati, Gen. Rel. Grav. **34** (2002) 1719.
[153] M. Consoli, Class. Quantum Grav. **26** (2009) 225008.
[154] G. Jannes and G. E. Volovik, JETP Lett.**96** (2012) 215.
[155] S. Finazzi, S. Liberati and L. Sindoni, Phys. Rev. Lett. **108** (2012) 071101.
[156] A. J. Leggett, Quantum liquids, Oxford University Press 2006, p.102.
[157] V. W. Hughes, H. G. Robinson, and V. Beltran-Lopez, Phys. Rev. Lett. **4** (1960) 342.
[158] R. W. P. Drever, Phil. Mag. **6** (1961) 683.
[159] C. M. Will, The Confrontation between General Relativity and Experiment, arXiv:gr-qc/0510072.
[160] U. Leonhardt and P. Piwnicki, Phys. Rev. **A60** (1999) 4301.
[161] J. M. Jauch and K. M. Watson, Phys. Rev. **74** (1948) 950.
[162] L. D. Landau and E. M. Lifshitz, Fluid Mechanics, Pergamon Press 1959, Chapt. III.
[163] J. C. H. Fung et al., J. Fluid Mech. **236** (1992) 281.
[164] L. A. Saul, Phys. Lett. **A 314** (2003) 472.
[165] E. Nelson, Phys. Rev. **150** (1966) 1079.
[166] P. Jizba and H. Kleinert, Phys. Rev. **D82** (2010) 085016.
[167] P. Jizba and F. Scardigli, Special Relativity induced by Granular Space, arXiV:1301.4091v2[hep-th].
[168] Ch. Eisele et al., Opt. Comm. **281** (2008) 1189.
[169] T. S. Jaseja, et al., Phys. Rev. **133** (1964) A1221.
[170] R. V. Pound, Rev. Sci. Instrum. **17** (1946) 490.
[171] R. W. P. Drever et al., Appl. Phys. B **31** (1983) 97.
[172] E. D. Black, Am. J. Phys. **69** (2001) 79.
[173] G. F. Smoot, M. V. Gorenstein, and R. A. Muller, Phys. Rev. Lett.**39** (1977) 898.
[174] H. Müller, et al. , Phys. Rev. Lett. **91** (2003) 020401.
[175] R. Schützhold, Class. Quantum Gravity **25** (2008) 114027.
[176] C. D. Hoyle et al., Phys. Rev. **D70** (2004) 042004.
[177] H. Yilmaz, Phys. Rev. **111** (1958) 1417.
[178] B. O. J. Tupper, N. Cimento **19B** (1974) 135; Lett. N. Cimento **14** (1974) 627.
[179] R. P. Feynman, in Superstrings: A Theory of Everything ?, P. C. W. Davies and J. Brown Eds., Cambridge University Press, 1997, pag. 201.
[180] R. D'E. Atkinson, Proc. R. Soc. **272** (1963) 60.

[181] K. Thorne, Black Holes and Time Warps: Einstein's Outrageous Legacy, W. W. Norton and Co. Inc, New York and London, 1994, see Chapt. 11 "What is Reality?".

[182] R. J. Cook, Am. J. Phys. **72** (2004) 214.

[183] A. S. Eddington, Space, Time and Gravitation, Cambridge University Press, 1920.

[184] A. M. Volkov, A. A. Izmest'ev, and G. V. Skrotski, Sov. Phys. JETP **32** (1971) 686.

[185] L. D. Landau and E. M. Lifshitz, The Classical Theory of Fields, Pergamon Press, 1971, p.257.

[186] J. Broekaert, Found. of Phys. **38** (2008) 409.

[187] M. Consoli and A. Pluchino, Eur. Phys. J. Plus **133** (2018) 295.

[188] D. A. Jennings et al. , Journ. of Res. Nat. Bur. Stand. **92** (1987) 11.

[189] K. Numata, A, Kemery and J. Camp, Phys. Rev. Lett. **93** (2004) 250602.

[190] J. A. Stone and A. Stejskal, Metrologia **41** (2004) 189.

[191] C. Lämmerzahl et al., Class. Quantum Gravity **18** (2001) 2499.

Index